Chromosomes

Chromosomes

The complex code

Melody **M.S. Clark**
Research Associate
Molecular Genetics Unit
University of Cambridge School of Clinical Medicine
Cambridge

and

W.J. Wall
Independent Consultant
London

CHAPMAN & HALL

London · Weinheim · New York · Tokyo · Melbourne · Madras

Published by Chapman & Hall, 2–6 Boundary Row, London SE1 8HN, UK

Chapman & Hall, 2–6 Boundary Row, London SE1 8HN, UK

Chapman & Hall GmbH, Pappelallee 3, 69469 Weinheim, Germany

Chapman & Hall USA, 115 Fifth Avenue, New York, NY 10003, USA

Chapman & Hall Japan, ITP-Japan, Kyowa Building, 3F, 2-2-1 Hirakawacho, Chiyoda-ku, Tokyo 102, Japan

Chapman & Hall Australia, 102 Dodds Street, South Melbourne, Victoria 3205, Australia

Chapman & Hall India, R. Seshadri, 32 Second Main Road, CIT East, Madras 600 035, India

First edition 1996

© 1996 M.S. Clark and W.J. Wall

Typeset in 12/14 Garamond by WestKey Limited, Falmouth, Cornwall

Printed in Great Britain by the Alden Press, Oxford

ISBN 0 412 75210 7 (HB)   0412 55530 1 (PB)

A catalogue record for this book is available from the British Library

Library of Congress Catalog Card Number: 98–83050

∞ Printed on permanent acid-free text paper, manufactured in accordance with ANSI/NISO Z39.48-1992 and ANSI/NISO Z39.48-1984 (Permanence of Paper).

Contents

Introduction: why study chromosomes?

Cytogenetics is probably one of the oldest branches of genetics. Humans have always been fascinated by heredity, primarily their own, but also that of their crops and animals. The use of chromosomes for prenatal diagnosis started about 20 years before the introduction of direct gene analysis and diagnosis of fetal chromosomal disorders is still of fundamental importance; banded chromosome analysis, introduced in the 1970s, increased the range of disorders that could be recognized.

Molecular biology is now fundamental in all areas of biological science, and cytogenetics is also of increasing value in the study of a wide range of genetic phenomena. The fundamental concept of looking at chromosomes under a microscope remains unchanged, but the methodology has dramatically altered over the years.

Gene expression, for example, is not entirely governed by a DNA sequence. The position of a gene in relation to a nucleosome and non-histone proteins also plays a role. These proteins and their three-dimensional conformation ultimately determine the accessibility of the transcription apparatus and therefore whether or not a gene is active.

Similarly, studies of gene organization have shown that the position of a gene on a chromosome is intrinsically linked to its expression. The combined effort of mapping and banding of chromosomes has shown that there is still much to be learned about the position of a gene on a chromosome. Questions of the higher order organization, such as 'why do certain genes seem to travel together?' and 'why does their position affect their expression?', remain unanswered.

The use of chromosomes, in the form of prometaphase spreads, or extended chromatin, has proved to be invaluable in gene mapping studies. This is a rapid and reliable method of ordering DNA sequences. By using these techniques of *in situ* hybridization biologists can physically map genes to chromosomes and also detect subtle changes in gene order. Alterations in gene order on a chromosome can be fundamental to the expression of oncogenes, which are pivotal in cancer genetics. Accurate delineation and prognosis of a cancer may depend upon careful cytogenetic analysis.

Cytogenetics is still of great importance when monitoring changes in somatic cell hybrids. The role of chromosome genetics within modern biology is continually changing and evolving, and thus a knowledge of cytogenetics is of ever-increasing value.

Chromatin structure and replication

1

Summary

It is now a well-known fact that DNA is the hereditary material and that the vast majority of the DNA of an organism is housed in organelles called chromosomes. However, the length of DNA contained in a cell is far greater than the size of the cell within which it exists; for example, the shortest human chromosome consists of 1.4 cm of DNA which must be compacted into a cell only a few micrometres across. To achieve this, DNA has to undergo several levels of packing, a process that is a mixture of DNA–DNA and DNA–protein interactions. The most compact form of DNA can be seen under the light microscope as a metaphase chromosome during mitotic cell division.

DNA must also be faithfully replicated to ensure accurate cell function and maintenance of the organism as well as ultimately, the reproduction of the species.

Our current knowledge of these complex processes is very limited, with huge gaps, in particular, concerning the links between the molecular and organelle level of organization. The current extent of this knowledge is examined in this chapter, although it has to be stressed that there are probably as many questions raised as answers given.

Box 1.1   Colchicine – The primary molecule of cytogenetics

Colchicine (Figure 1.1) was first used in cytogenetic studies during the 1950s. It was, without doubt, this molecule, more than any other, that was pivotal in allowing geneticists to visualize chromosomes during metaphase. Colchicine binds to the primary molecule of the cell spindle, tubulin. This occurs on a molecule-to-molecule basis.

Once inactivated by the binding of colchicine, tubulin cannot assemble into microtubules and therefore can no longer function as a control mechanism for chromosomes, with the net result that nuclear division is halted. This halting of cell division can be used to produce metaphase plates for later microscopic investigation. Alternatively, colchicine can be introduced into cell cultures and then washed out after an appropriate period. The net result of this is to produce polyploid cells, which can be a valuable tool in investigations of cell division.

Colchicine is an alkaloid derived from *Colchicum autumnale*, the autumn crocus, more specifically from the crocus bulb. This plant flowers in the autumn with a large pale mauve flower that is produced before any leaves appear above ground. Unusually, this plant provides another product which, unlike colchicine, a highly toxic product, is a spice of great value. This spice is saffron, which is derived from the stigmas of the plant.

It can be appreciated from this that autumn crocus bulbs, if eaten, are highly toxic, owing to interference with the cell cycle. It came as a surprise therefore to find that some rodents, such as the Chinese hamster, are able to eat *Colchicum* bulbs with no apparent ill-effects. When cells from these species were grown in culture it was found that it is not only the organism that is resistant; individual cells are highly resistant to colchicine; the dose required to stop cell division being very much higher than needed in other mammalian species.

Figure 1.1   Colchicine molecule.

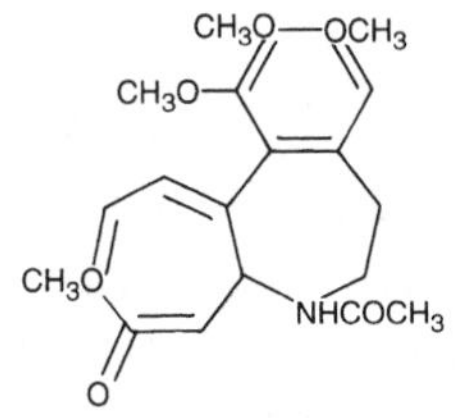

1.1   *Chromosome structure*

Chromosomes are visible under the light microscope only at certain stages of cell division. They are usually examined at mitotic metaphase when the chromosomes are maximally condensed. Cell division can be artificially stopped at this stage by the application of colchicine (Box 1.1).

When plain stained (Figure 1.2), chromosomes look like solid homogeneous structures. Very little detailed knowledge can be gained from such a cell. It is really only possible to count the total number of chromosomes and compare relative sizes and shapes (Figure 1.3).

Plain-stained chromosomes are very deceptive; their molecular complexity will be described below, starting with DNA itself. The nature of the individual structures such as centromeres and telomeres will be described in the next chapter. Methods by which individual chromosomes can be more accurately identified and the chromosomal composition of a cell defined will be described in Chapter 3.

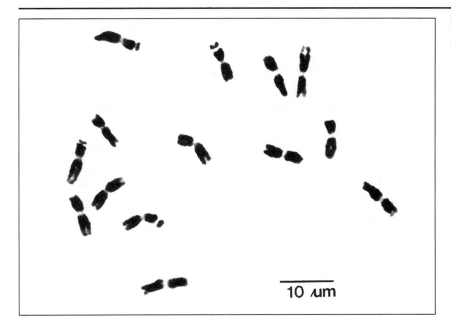

Figure 1.2 Metaphase spread of barley chromosomes, $2n = 2x = 14$. Plain stained with Giesma.

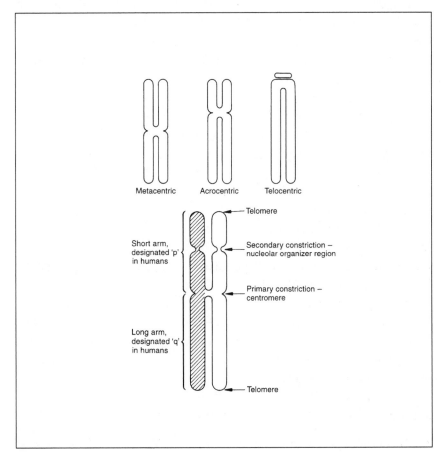

Figure 1.3 Terminology associated with describing chromosome structure.

1.2   Evidence that DNA is the genetic material

Nucleic acids were first discovered as long ago as the late 1860s by Friedrich Miescher. He isolated an unusual phosphorus-containing compound, which he called nuclein, from the nuclei of pus cells on discarded surgical bandages.

However, it was not until 1944 that it was proved that DNA codes for the cell's genetic information, when Avery, Macleod and McCarty expanded the previous experiments of Griffith in 1928. Griffith worked on pneumococcus strains of bacteria, of which there are many different types. Only one, *Streptococcus pneumoniae*, causes septicaemia (and death) in mice. This virulent strain can be distinguished from the others by the type of colonies it forms on agar plates. The virulent strain (designated S) produces smooth colonies owing to the presence of a polysaccharide coat that protects it against the host's defence mechanisms. The other non-virulent forms produce rough (R) colonies. Griffith found that mice injected with a mixture of heat-killed S cells and live R cells from pneumococcus died of septicaemia and that live S cells could be recovered from the dead mice. This virulence was found to be heritable, i.e. something that was heat stable in the S cells was transforming R cells into virulent strains.

Avery, Macleod and McCarty duplicated and refined this experiment *in vitro*. Live R cells were mixed with different purified components from heat-killed S cells: DNA, protein, capsule polysaccharides, etc. Only the DNA transformed the avirulent R cells into virulent S cells. Also, if the DNA from the heat-treated S cells was pretreated with DNAase (which digests DNA), this transformation effect no longer worked, i.e. DNA is the heritable material.

This discovery was greeted with great scepticism at the time, as there was still a strongly held belief that protein was the heritable material. Hence, further evidence was required to substantiate these experiments. Hershey and Chase in 1952 labelled T2 bacteriophage radioactively, in either the protein (with ^{35}S) or DNA (with ^{32}P) components. When the host bacterium (*Escherichia coli*) was incubated with these labelled bacteriophages, the viral DNA containing ^{32}P rapidly entered the host and replicated to form new virus. The protein (and ^{35}S) remained outside, producing no effect.

In the period 1949–53, Chargaff and colleagues made numerous important observations on the composition of DNA.

- It is composed of four bases – adenine (A), guanine (G), cytosine

(C) and thymine (T) – which vary in composition according to the organism (Table 1.1).

- DNA in different cells from the same organism has the same composition.
- The amount of DNA in cells from any species or organism is remarkably constant. It is not altered by external influences such as environmental factors, diet or metabolism.
- DNA from closely related organisms has a similar base composition, whereas distantly related organisms have widely differing DNA compositions.

In nearly all the DNAs isolated the number of adenine residues is always equal to the number of thymine residues and the number of cytosine residues equals the number of guanine residues (Chargaff's rule). The number of purines (A+G) is equal to the number of pyrimidines (T+C). These findings led to the acceptance of DNA as the genetic material. It is now known that RNA (ribonucleic acid) also codes for the genetic material in some viruses.

The finding of base equivalences raised the possibility of a structural organization to DNA. This led to the X-ray crystallography studies in the early 1950s of Wilkins and Franklin and to the postulation of the three-dimensional structure of DNA by Watson and Crick in 1953: the double helix. For further detail on the history of cytogenetics, refer to Chapter 11.

1.3 The double helix

DNA is a chain of nucleotides. Each nucleotide comprises a sugar (deoxyribose), a phosphate group and a base. This base is either a

Organism	G + C (%)
Slime mould	22
Sea urchin	35
Yeast	39
Man	40
Salmon	44
E. coli	51
T7 bacteriophage	51
Herpes simplex virus	72

Table 1.1 Percentage (G + C) content of DNA from a variety of organisms

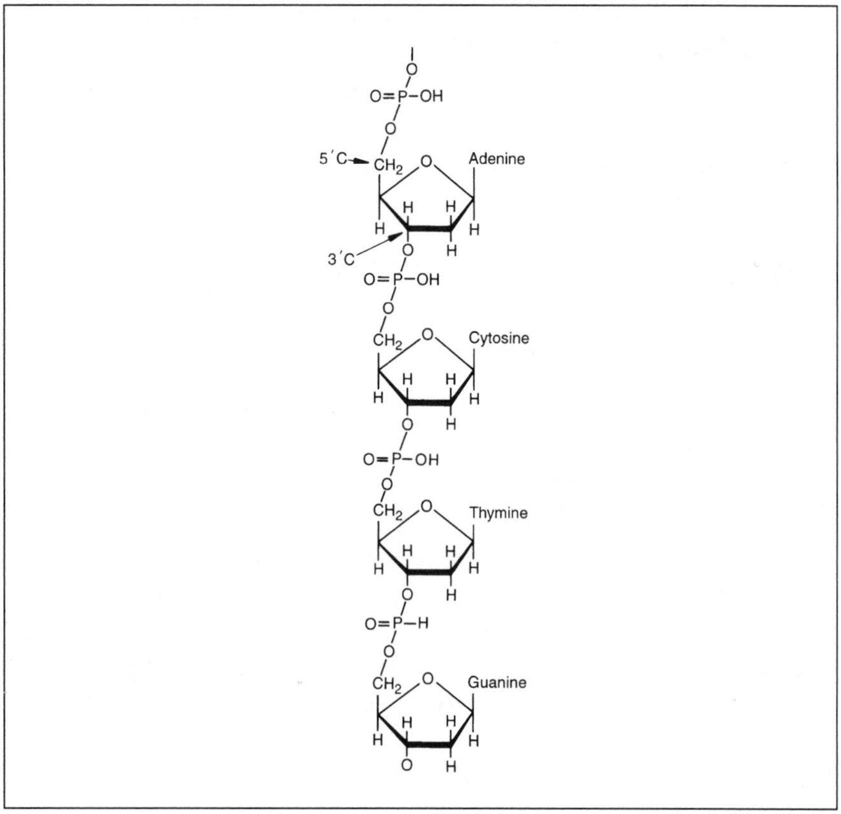

Figure 1.4 A section of DNA chain, showing the sugar–phosphate backbone. The bases are not drawn in full. Reproduced with permission from Adams *et al.* (1981) *Biochemistry of the Nucleic Acids*, Chapman & Hall.

purine (adenine or guanine) or a pyrimidine (cytosine or thymidine). The phosphates and sugars provide the external backbone structure with the bases projecting. A molecule of DNA is polarized 5′ to 3′

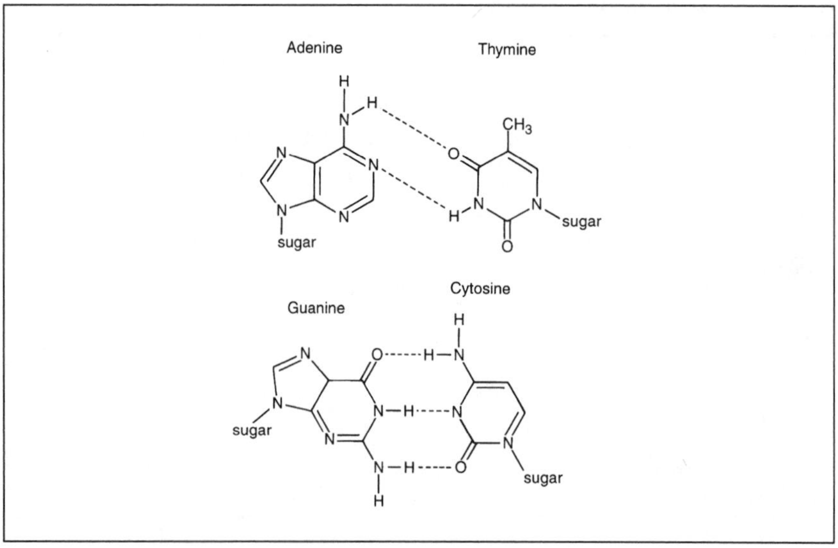

Figure 1.5 The normal base pairing found in DNA between adenine and thymine and between cytosine and guanine. Adenine and guanine are purines, cytosine and thymine are pyrimidines. Reproduced with permission from Adams *et al.* (1981) *Biochemistry of the Nucleic Acids*, Chapman & Hall.

as a result of the phosphate ester between the 3′ hydroxyl group on the sugar of one nucleotide and the 5′ phosphate group on the next (Figure 1.4).

To form the double helix proper, two strands of DNA pair in opposite polarities, with the phosphate–sugar backbone on the outside and bases projecting into the middle. The pairing attraction holds because of the formation of hydrogen bonds between the purines and pyrimidines. Because of their structure adenine always pairs with thymine and guanine with cytosine (Figure 1.5).

The conformation adopted by the double-stranded DNA molecule is generally that of a right-handed double helix, similar to a spiral staircase (Figure 1.6). The two strands cannot be separated without unwinding and present a very stable structure. There are 10 base pairs in each turn of the helix.

When writing out a DNA sequence, the sugars and phosphates are ignored and single letters are used for the bases, e.g. A (adenine), C (cytosine), T (thymidine) and G (guanine). Although DNA consists of a double-stranded molecule, because of the exact nature of the pairing arrangements it is considered unnecessary to write out both strands and a single strand only is usually written out. The bases (A,T,G,C) in a sequence are referred to as base pairs (bp), which is one of the measurements used when describing DNA (1000 bp = 1 kilobase pair, denoted kb; 1 000 000 bp = 1 megabase pair, denoted Mb).

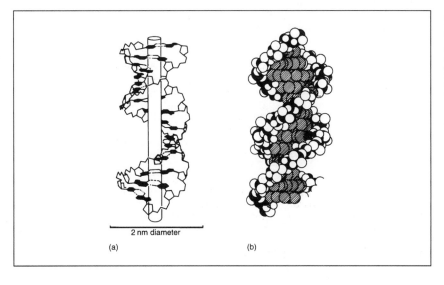

2 nm diameter

(a)   (b)

Figure 1.6 DNA double helix showing (a) helix and base pairs and (b) the more commonly recognized space-filling model. Reproduced with permission from Adams *et al.* (1981) *Biochemistry of the Nucleic Acids*, Chapman & Hall.

1.4.   The genetic code

DNA within the chromosomes acts as a template for transcription and translation into proteins. Transcription takes place in the nucleus, where an exact RNA copy (except that the thymine base is replaced by uracil) of the DNA template is produced. This RNA intermediate is called messenger RNA or mRNA. This is transported from the nucleus into the cytoplasm, where it is processed by the *in vivo* protein translation mechanisms to form proteins (chains of amino acids). The RNA (and therefore the DNA) sequence is a code for the production of a chain of amino acids.

The DNA bases form a 'commaless' code that is read in groups of three (called codons) by ribosomes, the cell's protein manufacturing machinery. There are 64 different combinations of triplets of four bases, allowing the possibility of coding for 64 different amino acids; however, the code is said to be degenerate since there are in fact only 20 different amino acids (Table 1.2).

Because of this degeneracy, the mutation of a single base pair (particularly if it is the third one in the codon) does not necessarily result in a change in the amino acid constitution of the protein. Three of the

Table 1.2   The genetic code

First base	Second base								Third base
	U		C		A		G		
U	UUU	Phenylalanine	UCU	Serine	UAU	Tyrosine	UGU	Cysteine	U
	UUC	Phenylalanine	UCC	Serine	UAC	Tyrosine	UGC	Cysteine	C
	UUA	Leucine	UCA	Serine	UAA	End	UGA	End	A
	UUG	Leucine	UCG	Serine	UAG	End	UGG	Tryptophan	G
C	CUU	Leucine	CCU	Proline	CAU	Histidine	CGU	Arginine	U
	CUC	Leucine	CCC	Proline	CAC	Histidine	CGC	Arginine	C
	CUA	Leucine	CCA	Proline	CAA	Glutamine	CGA	Arginine	A
	CUG	Leucine	CCG	Proline	CAG	Glutamine	CGG	Arginine	G
A	AUU	Isoleucine	ACU	Threonine	AAU	Asparagine	AGU	Serine	U
	AUC	Isoleucine	ACC	Threonine	AAC	Asparagine	AGC	Serine	C
	AUA	Isoleucine	ACA	Threonine	AAA	Lysine	AGA	Arginine	A
	AUG	Methionine	ACG	Threonine	AAG	Lysine	AGG	Arginine	G
G	GUU	Valine	GCU	Alanine	GAU	Aspartic acid	GGU	Glycine	U
	GUC	Valine	GCG	Alanine	GAC	Aspartic acid	GGC	Glycine	C
	GUA	Valine	GCA	Alanine	GAA	Glutamic acid	GGA	Glycine	A
	GUG	Valine	GCG	Alanine	GAG	Glutamic acid	GGG	Glycine	G

codons designate the end of a message (chain terminators) and one codon (AUG) for methionine acts as a start signal for protein synthesis. The intricacies of transcription and translation are beyond the scope of this book and the reader should refer to more advanced molecular texts.

Diseases are frequently caused by a mutation in a single base pair of DNA (e.g. sickle cell anaemia) which alters only a single amino acid. The properties of the changed amino acid may have severe consequences for the individual as it is the interactions between the amino acids in the protein chain that cause the protein to fold and adopt its three-dimensional conformation. Alterations in the protein structure may destroy or weaken the action of the protein, leading to metabolic deficiencies. Diagnosis of the causes of genetic disease has only recently been possible in the light of our ability to sequence and understand the properties of DNA and proteins.

DNA, although coding for proteins, also has encrypted within it all the necessary control sequences and regulators required for the correct expression and production of these proteins and ultimately the successful maintenance of the organism. Diseases may be caused by mutations in these control sequences as well as by changes in protein structure. DNA sequences have also been implicated in the higher order organization of the chromosome. It is important to appreciate that DNA affects more than just the protein it encodes.

1.5 DNA constitution

As demonstrated by Chargaff, DNA, in both the form of base composition and the type of DNA, varies from organism to organism.

DNA sequences can be divided into three types depending on the number of times a sequence appears in the genome.

1. Low- or single-copy DNA. These are sequences encoding most enzyme functions. In general, these can constitute up to 50% of the total DNA.
2. Middle repeat DNA. These sequences code for most of the structural components of a cell, such as the histones, ribosomal RNA and transfer RNAs(tRNAs). This class can also contain retrotransposons or retrovirus-like sequences (which are assumed to be inactive and redundant) such as the Alu repeats in humans. Thirty to forty per cent of the genome may comprise middle repeat DNA.

3. Highly repetitious DNA. This is simple sequence DNA and is frequently non-coding. In mammals, this can constitute 20–50% of the genome, but in many plants this figure can exceed 80%. This type of DNA often has a different constitution to the rest of the DNA and can be isolated on density gradients as satellite DNA. It is frequently associated with specific chromosome structures, such as heterochromatin (see Chapter 2 for more detail).

In general, DNA in the genome is arranged with single-copy sequences interspersed with either repetitive or middle repeat DNA.

The complexity of the genome, i.e. the proportions of each type of sequence, is determined using Cot curves (Box 1.2).

Box 1.2　Cot curves

The calculations for working out Cot curves are complex, but fortunately the results are invariably presented in a uniform way and are easily interpreted.

Cot is a modification of $C_o \times t$, or initial concentration of DNA in moles of nucleotide per litre (C_o) × renaturation time in seconds (t). The values are plotted on a log scale, resulting in a curved graph whose slope gives an indication of the degree of repetition of the sequences in the organism (Figure 1.7).

The Cot technique is based on the observation that extremes of temperature and pH can dissociate the double helix into two separate strands. Given suitable conditions these strands will reanneal with their complementary sequences to form stable double helices. If the DNA is sheared into small, i.e. 400–600 bp pieces, and denatured, the highly repetitious sequences will reanneal first as they have more chance of finding a complementary sequence. The low-copy or single-copy sequences will anneal last, as they take longer to find their opposite strand. The more complex the genome, the longer these sequences will take to find their complementary strand.

At various time intervals after shearing and denaturing, the DNA solution that is being gently annealed is passed through a hydroxyapatite column (which binds DNA). The DNA is then released from the column using different salt solutions to alternatively release the single- and the double-stranded DNA. The percentage of annealed DNA can then be determined using an optical densitometer.

Figure 1.7 A Cot curve. DNA reassociating at low Cot values (10^{-4} to 10^{-1}) is composed of highly repetitive DNA. Cot values of 10–100 are moderately repetitious and DNA annealing at higher Cot values is non-repetitive (low or single copy).

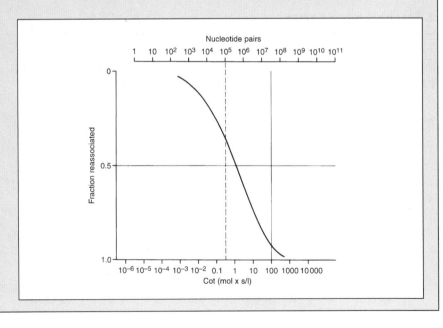

Table 1.3 DNA content in a variety of eukaryotes

Species	Common name	1C nuclear DNA content(pg)	n
Fritillaria davisii	Lily species	98.4	12
Protopterus	Lungfish	50	19
Lolium longiforum Thumb.	Lily species	35.2	12
Avena sativa	Oat	21.5	21
Triticum aestivum	Bread wheat	18.1	21
Allium cepa	Onion	16.8	8
Homo sapiens	Man	3.7	23
Mus musculus	Mouse	2.5	10
Drosophila	Fruit fly	0.1	4
Arabidopsis thaliana	Annual weed	0.07	5
Saccharomyces cerevisiae	Yeast	0.026	15

The amount and composition of DNA does not necessarily reflect the complexity of an organism (Table 1.3).

The fact that the DNA content of an organism is much greater than that required to code for and regulate the production of all necessary proteins has been termed the C-value paradox.

In eukaryotes the C-value is defined as the amount of DNA per genome, as follows:

1C: haploid nuclei, i.e. egg and sperm;
2C: diploid nuclei;
4C: nuclei which are just about to divide by mitosis.

This excess DNA is mainly present in the form of highly repetitious sequences. In general, the smaller the haploid genome, the smaller the amount of repetitive DNA i.e. *Arabidopsis* (which has the smallest genome size found in flowering plants) has 7×10^4 kb of DNA with only 14% in the form of highly repetitious sequences, whereas maize with 15×10^6 kb has almost 80% of its DNA in this form.

The reason for calling this finding a paradox is fairly obvious: no-one has yet determined why organisms need these extra sequences. However, it has been shown that nuclear DNA amount affects cell growth, development and cell division, aspects of which are particularly well studied in plants. While this may be an adaptive characteristic, to a certain extent, this cannot represent the whole story. Because of its mainly non-coding nature, highly repetitious DNA is often referred to in books as 'junk' DNA. There has been (and will continue to be) a long-running debate as to its use. It is becoming clear that some repetitive sequences, such as the human Alu repeat family, occur in particular chromosomal domains (discussed further in Chapter 3) and may well play a structural role within the chromosome. It is also thought that this 'extra' DNA provides the genetic material on which mutation and evolution can act to produce new genes or adaptations to new environments.

This vast amount of DNA in the form of a double helix is then complexed with numerous proteins and undergoes extensive packing to form the metaphase chromosome structure.

1.6   Higher order structure of DNA

Higher order structure can be divided into two types.

1.   Prokaryotes, e.g. viruses and bacteria. These organisms generally contain one-hundredth of the DNA of eukaryotes. The DNA is thinly coated with proteins that are involved in the folding processes and also the enzymes required for repair, recombination and transcription. Some viruses contain RNA as the genetic material. These represent much simplified structures when compared with eukaryotic chromosomes and will not be discussed further. Allied to prokaryotic DNA structures are two types of extranuclear DNA in eukaryotes, that found in the mitochondria and chloroplasts (the latter is found only in plants). These reside in the cytoplasm and are replicated and expressed by their own specific genetic apparatus. They carry important genes involved in the respiratory pathways and also ribosomal and tRNAs. Their products interact with those of the nuclear genes and are essential for the normal functioning of the organism. These DNA molecules are more similar to the prokaryotic type, being circular, double-stranded, supercoiled molecules, and as such will not be discussed here.

2.   Eukaryotes. Eukaryotic chromosomes are encased in proteins, in a 2:1 ratio of protein to DNA. These proteins consist of the five major histones plus a very heterogeneous mix of 30 non-histone proteins (called HMGs, high-mobility group proteins) and some associated RNA. This mix as a whole is termed chromatin. Each human chromosome comprises between 1.4 and 7.3 cm of DNA, packaged into a few micrometres of chromatin. This constitutes a packing ratio of 1:10 000, most of which is due to the interaction of the chromatin proteins.

It is generally agreed that chromatin is organized on four basic levels, as follows:

- primary: nucleosomes;
- secondary: solenoids;
- tertiary: loops;
- quarternary: final folding into chromosome shape.

The first two stages are well established. However, the others are poorly understood and still controversial.

1.6.1   Primary structure: the nucleosome

This is the fundamental unit of chromatin. When DNA is subjected to an osmotic shock it unwinds to give a beaded string appearance, approximately 10 nm in diameter.

The beads are entirely composed of histone proteins. These proteins are almost universally associated with nuclear DNA in eukaryotes and represent some of the most evolutionarily conserved proteins. Each histone contains a hydrophobic core region with one or two basic arms. They contain an unusually high proportion of basic amino acids, which makes for strong interactions with DNA, which is acidic. They comprise five classes designated H1, H2A, H2B, H3 and H4, which combine to form a core complex comprising two molecules each of H2A, H2B, H3 and H4 and one molecule of H1. These histones may be modified by acetylation, methylation, phosphorylation, etc. Some of these changes alter the charge on the molecule and may affect the interactions between the histones and the DNA. For example, acetylation of H4 causes unfolding of the nucleosome core and is associated with transcription-ally active regions of chromatin, and dephosphorylation of H1 is associated with chromosome condensation.

If the beaded string is subsequently treated with trypsin, to remove H1, no difference in structure is observed, hence H1 is not vital to the beaded structure.

The histone complex can be dissociated into multimers, particularly an H3–H4 tetramer. Reassociation studies show that equimolar concentrations of H3 and H4 can spontaneously form tetramers in solution. In the presence of DNA the H3–H4 tetramer can bind two molecules each of H2A and H2B to form a DNA-associated octamer. This attaches periodically along any type of DNA irrespective of sequence and is called the nucleosome. H1 is not essential to this process.

DNA within the beaded structure is susceptible to endolytic digestion, suggesting that the DNA is wound around the outside of the nucleosome octamer, with H1 acting as an anchor. The rate of DNA

Figure 1.8 First level of DNA compaction: the 'beaded string' structure. See text for details.

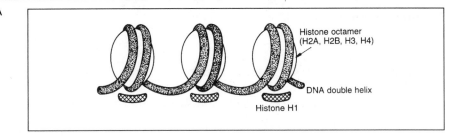

winding is 1.7 turns per core particle, producing a six- to sevenfold reduction in the length of the DNA. H1 is then bound at the edge where the DNA winds round the histones so that it can seal off the nucleosome structure completing two full turns of the DNA (Figure 1.8).

In terms of base pairs, the DNA associated with the nucleosome is organized into 200-bp units. This can be demonstrated by subjecting the nucleosome string to digestion by micrococcal nuclease (Figure 1.9). Size can be determined by running the digested fragments out on a gel. At an incubation temperature of 0°C, 200-bp fragments are produced with only one nick introduced per nucleosome. At higher temperatures, digestion proceeds as far as is possible until the DNA is protected by the histones. At 37°C 145-bp and 168-bp fragments are produced. Protein analysis of these fragments shows that in the 145-bp fragments only histones H2A, H2B, H3 and H4 are present. Within the 168-bp

Figure 1.9 Digestion of nucleosomes with micrococcal nuclease.

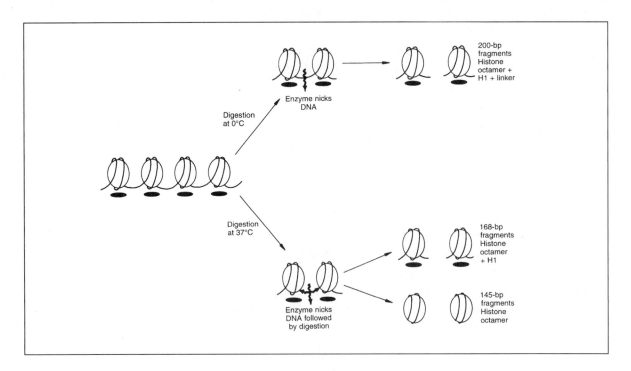

fragments, H1 is also present protecting an extra 23-bp of DNA. In conclusion, 145-bp are associated with the histone core particle. A further 23-bp are protected by histone H1. The remaining 60-bp form a linker between the adjacent nucleosome cores.

The histones *in vivo* have been shown to be associated with a variety of proteins, e.g. N1, N2, nucleoplasmin, which have been demonstrated to be necessary for nucleosome assembly. These are acidic proteins that act as molecular chaperones, neutralizing electrostatic repulsion between histones and promoting histone–histone interactions. Being acidic, they counteract the strong electrostatic attraction between histones and DNA, minimizing non-specific aggregation and precipitation.

1.6.2 Secondary structures: the solenoid

The nucleosome is supercoiled and organized into a solenoid structure, with 6–7 nucleosomes per turn (Figure 1.10). This is stabilized by interactions involving the highly variable C- and N-terminal regions of the H1 histone and also possibly the HMGs. This supercoiling produces a fibre of approximately 30 nm in diameter.

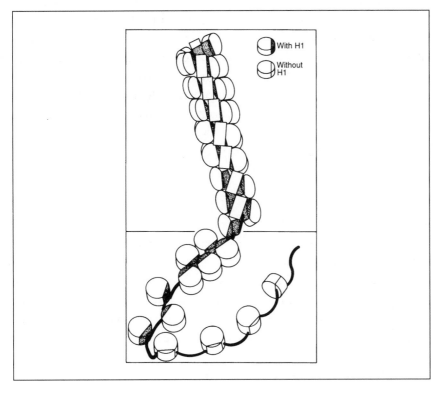

Figure 1.10 Second level of chromatin organization: the solenoid.

Figure 1.11 The proposed loop organization of DNA. The genomic DNA is organized into loops of between 5 and 200 kb. The loops are anchored to the protein scaffold at very specific DNA regions called SARs. Loops can have several transcriptional units, indicated by large open arrows. Reproduced with permission from Gasser, S.M. and Laemmli, U.K. (1987) *Trends in Genetics*, **3**, 21, Fig. 5.

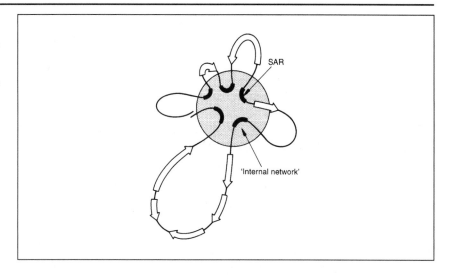

1.6.3 Tertiary structure: loop structure

When metaphase chromosomes are entirely depleted of histones by extraction with 2 M NaCl, the residue is an axial fibrous network (a protein scaffold) surrounded by a halo of DNA fibres which are clearly organized into loops that radiate in all directions (Figure 1.11). The fibres of the loops comprise the solenoid structure, described above. The bases of the loops merge into the logitudinal axis of the chromosome. These loops contain between 5 and 200 kb of DNA, with an average size of 63 kb. The loops are held together at their bases by non-histone proteins. In the condensed metaphase chromosome, it is proposed that neighbouring loops are held together by protein–protein or protein–DNA interactions, forming an internal network along the chromosome axis. It is suggested that the loops represent a functional subunit of the chromosomes, containing different sets of functionally linked genes corresponding to replicons – the units of replication.

What defines a DNA loop? If the scaffold structure is subjected to restriction enzyme digestion, characteristic DNA sequence fragments co-sediment with the nuclear scaffold. These have been termed scaffold-associated regions (SARs) or occasionally matrix-associated regions (MARs). SARs (Box 1.3) define loops of various sizes and are usually found in non-transcribed regions. Each loop can contain one or more genes, and it is suggested that the size of the loop is correlated to transcriptional activity, the smaller loops generally containing highly transcribed genes.

Recent evidence suggests that the loops are in a dynamic state in

---

Box 1.3   SARs

SARs are located in AT-rich regions and three DNA sequence motifs have been identified, as a weak consensus sequence:

- T box: [TT(A/T)T(T/A)TT(T/A)TT];
- A box: [AATAAA(T/C)AAA];
- Topo II cleavage site: [GTN(A/T)A(T/C)ATTNATNN(G/A)].

There are usually eight or more elements related to the topo II (DNA topoisomerase II) cleavage consensus sequence plus several A and T boxes forming an AT-rich region. Topo II has a major role as a SAR-binding protein. It has been found that topo II preferentially binds and aggregates SAR-containing fragments, a process probably relevant to chromosome condensation.

It is thought that SAR binding to the scaffold is not entirely sequence-specific, but more a function of DNA shape and the ease with which the sequence unwinds.

---

relation to the scaffold. Analysis of specific DNA sequences found within the loops indicates that transcriptionally active DNA is tightly condensed and associated with the scaffold. Transcriptionally inactive sequences are in a more extended form and occur within the loops themselves. As not all cells have the same set of functional genes, this implies that rearrangements in chromatin packaging occur in relation to cell cycle and gene expression. Each gene, therefore, does not have a permanent fixed position with regard to the scaffold structure, although there must be a higher order defined, perhaps imposed by SARs, to result in the final metaphase chromosome structure.

1.6.4   Quarternary structure: final folding into chromosome shape

This involves the scaffold structure (Figure 1.12) briefly mentioned above. It can be isolated as an independent structure and has the size and shape of an untreated chromosome. It contains no histones, but a mix of 30 non-histone proteins (the HMGs). The major proteins of this highly variable class have been designated Sc1 and Sc2. Sc2 has a molecular mass of 135 kDa and, as yet, no role has been assigned to it. However, Sc1 (molecular mass 170 kDa), the most abundant, has been identified as the enzyme DNA topoisomerase II (topo II).

Antibodies raised against topo II and hybridized to the scaffold outline the whole structure. It has been shown that topo II is present in approximately three copies per DNA loop, and this is consistent with its proposed role as a loop fastener. Its structural and enzymic role is highly regulated and may be involved in the release of stress during transcription and replication. It is positioned at the base of the loops and hence is strategically situated to control long-range order in

Figure 1.12 Electron micrograph of a histone-depleted metaphase chromosome showing the central protein matrix (scaffold). Note that the characteristic shape of the chromosome is maintained. Reproduced with permission from Laemmli *et al.* (1977) *Cold Spring Harbor Symposia on Quantitative Biology XLII*, 355.

chromatin, and it is hypothesized that topo II plays an important role in chromatin condensation.

Several models have been proposed for the further folding of DNA after the loop structures, all containing the following basic elements (Figure 1.13).

The loops (described above) radiate from a central protein core that follows a helical path along the chromatid axis of the chromosome. These loops are then arranged into hexameric rosettes, comprising a total of 300 kb of DNA, producing a 200–300 nm fibre. This is the form of the decondensed interphase chromosome. These loops are further arranged

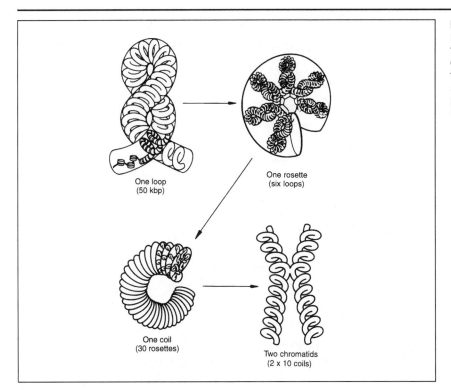

One loop
(50 kbp)

One rosette
(six loops)

One coil
(30 rosettes)

Two chromatids
(2 x 10 coils)

Figure 1.13 A proposed model for the higher order folding of chromatin into chromosomes. Reproduced from Filipski *et al.* (1990), *EMBO*, **9(4)**, 1325, with permission of Oxford University Press.

into coils (approximately 30 rosettes per coil), with each coil equivalent to an average human G-band (approximately 9 Mb of DNA) (for details of G-banding, see Chapter 3). These coils are then further condensed to produce the final metaphase chromosome structure. The number of coils is dependent upon the amount of DNA in each chromosome; it is estimated that human chromosome 1 consists of 29–33 coils.

The means by which this tertiary and quarternary folding is achieved is still poorly understood. However, several proteins which have been implicated as playing a role in this process have been isolated. One of these, AF-2, is a novel nuclear antigen, the presence of which has been shown to be related to cell cycle-dependent alterations in chromatin structure. A more extensive group of proteins, the SMC family, has been discovered. Ten members are known so far, isolated from diverse origins such as yeast, *Xenopus, Caenorhabditis elegans* and mouse. These show structural similarities to motor proteins, are scaffold associated and have been demonstrated to play an important role in mitotic chromosome condensation. It is hypothesized that these are novel motor proteins that facilitate the formation of chromosome loops, perhaps interacting with the role of topo II.

Evidence is rapidly accumulating to support the existence and function of SARs, but the further levels of folding and the cell cycle dynamics of its structure are still ill-defined and subject to debate.

1.7   Chromosome replication

Closely allied with the definition of chromatin structure is the process of replication, whereby the relatively huge molecular structure of a chromosome is duplicated at the DNA level and reformed into two separate organelles.

This process is highly regulated in both prokaryotes and eukaryotes to ensure that the genome is accurately duplicated each time. In order to effect this, chromosomes are replicated semiconservatively, i.e. each strand of DNA is used as a template for the synthesis of a duplicate. The use of a template strand ensures that a correct copy of all the sequences is present which is available as a check against accuracy, in other words a proofreading mechanism.

This semiconservative mode of replication was first suggested by Watson and Crick in 1953, after they described the double helical structure of DNA. However, it was not until 1957 that Taylor and colleagues conclusively proved this with their experiments on *Vicia faba* (Figure 1.14).

Seedlings of *Vicia faba* were grown in tritiated thymidine for 8 h, i.e. a full replication cycle. Any DNA synthesized incorporated the label

Figure 1.14   Taylor's experiment to demonstrate the semiconservative nature of DNA replication. The dark regions are those which have incorporated radioactive thymidine. See text for details.

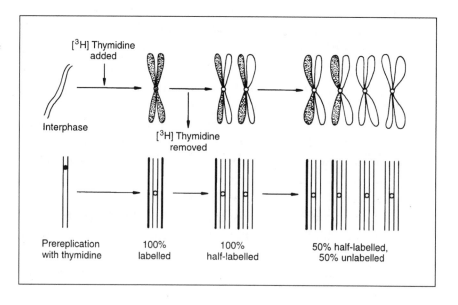

within the chromosomes. The roots were then transferred to a normal growth medium to undergo further division cycles. This medium also contained colchicine so that the spindle apparatus was destroyed. The result of this was that the chromosomes still replicated but were unable to partition between cells, so chromosomes from successive rounds of nuclear division remained in the same cell.

In this way it was possible, by counting the number of chromosomes present in a cell, to determine how many rounds of division had taken place. Slide preparations were made of the roots. These were then covered with a thin layer of photographic emulsion. The radioactive thymidine exposed spots on the emulsion, so that after the slides were developed and the chromosomes stained, spots appeared on the chromosomes where the thymidine had been incorporated and indicated the fate of the original labelled material.

The slides showed that initially the whole chromosome was labelled. The first division after the removal of the isotope showed all chromosomes to be labelled on one half only. In the next division, half the chromosomes were totally unlabelled while the others were half-labelled.

The only explanation was as follows. Before replication the chromosomes consisted of a single double helix molecule of DNA, i.e. containing two strands of DNA that were both labelled. On replication, one-half of the helix was passed onto each daughter chromatid. The initial labelling with tritiated thymidine only labelled one strand of each chromatid, but the resolution of the technique was limited in that it could not differentiate between labelled and unlabelled DNA strands within the same chromatid, hence it appeared that the whole chromosome was labelled. When these chromosomes replicated in the absence of the isotope, each metaphase chromosome formed one chromatid labelled on one strand only and one unlabelled chromatid. When these divided further, the pattern was repeated, with one chromosome being half-labelled and all others having no label at all, i.e. the chromosome was replicating semiconservatively. This also proved conclusively that the unreplicated chromosome contained one long continuous fibre of DNA (Box 1.4).

Experiments similar to those performed by Taylor *et al.* measured the length of cell division cycles of eukaryotes and compared the results with those obtained from *E. coli*. It was soon realized that, assuming that DNA replication rates are similar in all organisms, for chromosome replication to take place in the observed amount of time, replication would have to proceed from several points of origin. This was clearly demonstrated by Huberman and Riggs in 1968.

They adapted an autoradiographic technique designed to observe the replication of the *E. coli* chromosome to mammalian chromosomes. Their observations yielded the following information.

1. Replication forks move at about 3 kb per minute.
2. Each chromatin fibre contains numerous origins which are spaced on average 100 kb apart with a range of 15–300 kb.
3. These may be activated coordinately in blocks of 5–10.
4. Most replication forks occur as divergent pairs, indicating that they arise from bidirectional origins of replication (Figure 1.15).

Highly conserved sequences that act as origins of replication have been found in bacteria and also yeast. The average size of a replication unit as found by Huberman and Riggs implies that there are 20 000–50 000 origins of replication in the mammalian genome. However, despite this huge number, sequences similar to those found in bacteria and yeast have not been discovered in man. It is now thought that initiation of replication in most eukaryotes occurs at zones rather than at specific sites. These 'zones' have been examined in four eukaryotic genes: *Drosophila* chorion, human c-*myc*, Chinese hamster ovary (CHO) dihydrofolate reductase and CHO rhodopsin. They revealed four elements in common (Box 1.5).

These zones contain a number of sequences that are present throughout the genome. However, the particular grouping described

Figure 1.15 Chromatin bidirectional origins of replication.

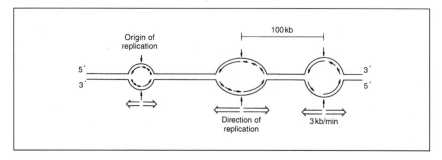

Box 1.5   Replication initiation elements
1. DNA unwinding elements (DUEs). These are inherently unstable duplex DNA segments characterized by their hypersensitivity to single-strand specific nucleases. No specific base sequence is responsible, but regions can be identified using computer programs that calculate nearest-neighbour free energy values.
2. Pyrimidine tracts (cytosine or thymine). At least one perfect pyrimidine tract (more than 12 adjacent pyrimidines) was found flanking each DUE.
3. Scaffold associated regions (SARs). These are consensus sequences that have been previously described.
4. Transcription binding factors. Interaction between transcription and DNA replication occurs in both prokaryotes and eukaryotes. Several proteins that function as transcriptional activators or repressors appear to play an important role in DNA replication.

---

above ensures a high probability that they will act as initiators of replication. The apparent complexity of replication origins in eukaryotes compared with prokaryotes is proposed as a natural consequence of the vastly larger genome and simply reflects more complex requirements for regulation rather than differences in basic mechanisms. It is thought that activation of an origin of replication is more reliant on position, i.e. accessibility within the three-dimensional structure of the chromosome, rather than specific sequence organization.

Once the origin is activated, the DNA is unwound from the double helix and single-stranded sections exposed and stabilized. When these are of a sufficient size the replication machinery kicks into action, producing an exact copy of each strand.

A clue as to where the replication machinery is active was provided by bromodeoxyuridine (BrdU) incorporation experiments into diffuse DNA loops. Chromosomes were treated with salt to extract the histones and other proteins. The DNA was then distended into loops surrounding the central scaffold structure to produce a nuclear halo. Preparations were made from cells which had been grown in BrdU for 15 mins followed by chase periods (in non-BrdU media) of varying durations ranging from zero to 18 h.

After short chase times discrete BrdU staining was found at sites close to the scaffold. With longer chase periods, the label progressively moved out into the extended DNA loops or halo. This implies that the sites of the replication machinery complexes are fixed at or near the chromosome scaffold and that the DNA is in a constant flux as it moves relative to the replication complex.

Once initiated, replication proceeds via the action of DNA polymerase I. This joins nucleotides together to form a new DNA chain. Synthesis of DNA occurs 5′ to 3′; however, with a bidirectional model, one of the strands would have to elongate in the 3′ to 5′ direction. Enzymes which allow this have not been found. Okazaki in 1968 came

Figure 1.16 Replication of DNA can only proceed in the 5′ to 3′ direction. One strand can be continually synthesized in the 5′ to 3′ direction. The other (lagging) strand is synthesized in pieces (Okazaki fragments), which are then ligated together with DNA ligase to form a complete DNA chain.

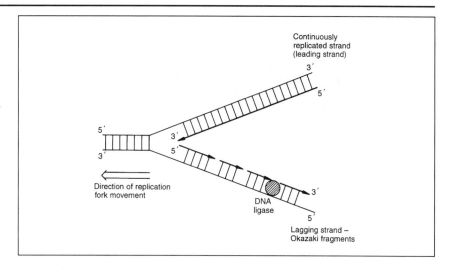

up with the solution that one strand is made continually in the 5′ to 3′ direction while the other is made backwards in short segments with DNA ligase joining the gaps (Figure 1.16). This was proved by his work on *E. coli* and T4 bacteriophage. During DNA synthesis these organisms were briefly pulsed with tritiated thymidine. When the DNA was denatured and isolated, very short pieces of labelled DNA were recovered. These short stretches of DNA are now known as Okazaki fragments.

DNA polymerase I does not act alone, but works in tandem with other proteins that increase the fidelity of its copying mechanism. It also requires a short stretch of nucleotides at the 5′ end to act as a primer to start off the reaction. Very short RNA sequences (about 10 bp long in animals) have been found covalently associated with Okazaki fragments. These do not have a specific base sequence and, once used as primers, are removed by an exonuclease. Other enzymes such as helicases and topoisomerases unwind the DNA at the replication fork and also release the tension within the DNA molecule when the double helix itself is unwound. Once this process is completed, those single strands which are temporarily exposed before replication begins are stabilized by a DNA-binding protein. A primase–polymerase complex initiates RNA-primed DNA synthesis on the leading strand. After initiation, this function is taken over by another type of polymerase complex, allowing the primase–polymerase complex to become exclusively involved with the Okazaki fragments. The basic details are demonstrated in Figure 1.17.

Replication involving the ends of chromosomes is slightly more specialized than the explanation above, (see section 2.3 for more details).

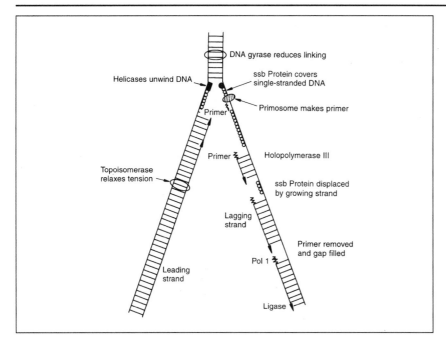

Figure 1.17 The major enzymes and proteins involved in DNA replication at the growing fork. Reproduced with permission from Adams *et al.* (1992) *Biochemistry of the Nucleic Acids*, Chapman & Hall.

The molecular reactions have been described very briefly; however, it must not be forgotten that this whole process takes place within the chromosome superstructure. The replication machinery must be able by some means to pass through the nucleosome-bound DNA and after passage reassemble the daughter strands into chromatin, re-establishing the higher order structure with the production of two separate daughter chromatids.

Investigations are under way to establish the fate of nucleosomes during replication. There is evidence that nucleosome dissolution to H3–H4 and H2A–H2B tetramers occurs during replication. Newly synthesized H3–H4 tetramers are selectively deposited on newly replicated chromatin, but old and new H2A–H2B histones are randomly associated as dimers with the new as well as the old H3–H4 tetramers. Evidence is accumulating for random segregation of nucleosomes between old and new DNA strands, however little is known about how the nucleosomes find their precise position on the two daughter strands.

The nucleosomes are low down the organizational level of the chromosome, and the role of higher order structure, including the non-histone proteins, has not yet been addressed. Unfortunately, knowledge concerning the mechanics of DNA replication has come from a mix of organelle-level experiments (Taylor) and molecular detail

(DNA–enzyme interactions); very little is known of the processes in between. These puzzles will be solved with time; however, what is becoming apparent is that, although the chromosome is viewed as a solid structure, its components are in dynamic equilibrium with their cellular environment.

Chromosome form and function

2

Summary
This chapter will deal with the structures found within chromosomes and the definition of chromosome parts. These include euchromatin, heterochromatin, centromeres and kinetochores, nucleolar organizing regions and telomeres. It is not always easy to define some of these structures other than as functional units since the definition intrinsically involves a simple description of the appearance of the structure as visualized by histochemical techniques.

We will also be looking at some of the unusual forms of chromosomes, such as polytene and lampbrush chromosomes, and double-minute and B chromosomes. Artificial chromosomes will also be dealt with in this chapter.

To start with we will give brief definitions of the substructures that are found within chromosomes and then go on to a more detailed and systematic look at them one at a time. The major physical structures are shown in Figure 2.1.

2.1 Structures in a chromosome

Chromosomes are constructed of two broad classes of DNA – euchromatin and heterochromatin – but within the chromosome there are also recognizable substructures. Euchromatin is really only a functional description; euchromatin is that part of DNA that is not heterochromatin. It is usually assumed that euchromatin is genetically

Figure 2.1 Basic structures found in a chromosome.

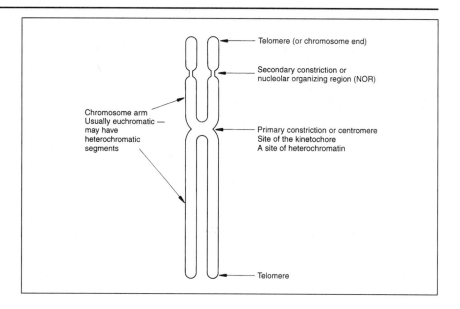

active and less contracted than heterochromatin although, because of great variation in euchromatin contraction, this distinction is difficult to define. Although heterochromatin is enriched with repeated sequences and highly contracted, we shall see that repeat sequences are frequently very important to the functioning of chromosomes. Heterochromatin also tends to be transcriptionally inactive, replicating late in S-phase.

The presence of repeat sequences does not in itself mean that a structure is inert. Consider, for example, the end caps of chromosomes. These are called telomeres. These vital structures stop the progressive shortening of chromosomes that would otherwise occur with each round of cell division. Although there are some notable exceptions, the repeat sequence of telomeres (TTAGGG)$_n$ is highly conserved across phylogenies.

Repeats of a different kind are also found at chromosome centromeres. The centromere is permanently contracted and therefore fits in with our functional definition of heterochromatin. In this particular case, however, the centromere is still an active feature, being the site of the kinetochore. This part of the chromosome is the point of attachment of the spindle fibres, which are important for chromosome separation during nuclear division.

Nucleolar organizing regions (NORs) also contain repeats, but in this case they are multiple copies of ribosomal RNA genes. NORs are unusual in often being associated with a specific area of the nucleus, creating a visible structure, the nucleolus.

There are some unusual forms of chromosomes which are sufficiently different from most types to warrant separate examination.

Lampbrush and polytene chromosomes are features of cells that are both species and tissue specific. In the case of polytene chromosomes a process of endoreduplication creates these enormous structures, essentially multiple repeats of the entire chromosomes. Although the reason for such overproduction in some tissues remains controversial, it is probably associated with the need for very high transcription rates.

In the case of supernumerary segments and chromosomes, the replication of sections, or even entire chromosomes, can have profound affects on the host organism. Double minutes and homogeneously staining regions (HSRs), for example, although normally small and inert, can contain multiple copies of one or two single genes that alter the metabolic abilities of the cells in which they are found.

Double minutes and HSRs are not regarded as normal additions to the cell, being induced in response to environmental stress. We can say that they are individual specific, whereas B chromosomes and supernumerary segments are generally uniform in structure within a population. They are regarded as a normal part of the chromosome complement of certain species. However, not all populations or individuals will contain B chromosomes or supernumerary segments, even if they are found in that species.

Using this knowledge of chromosome structure, it has recently become possible to construct totally artificial chromosomes. These represent new methods of investigating expression of specific individual genes and cloning specific sequences.

2.2 Heterochromatin

The silent nature of heterochromatin in transcription and translation is central to the interest that this material generates among geneticists. Heterochromatin is made up of repeat sequences that vary in length but tend to be short. The many attempts to relate heterochromatin content to specific clinical features in humans have all led nowhere. By contrast, the variation in heterochromatin in animal species has been shown to be important in speciation.

Heterochromatin has caused much trouble to geneticists in the past, not least because of the shifting sands of definition. Heterochromatin has in the past been thought of as the chromatin that can be delineated by the process of C-banding. As a working hypothesis this can be regarded as reasonable, but it must be remembered that inactive and condensed euchromatin is not the same as heterochromatin. We can say that constitutive heterochromatin is structurally and chemically distinct from euchromatin and that what is sometimes referred to as facultative heterochromatin is not heterochromatin at all. On this basis the Facultative heterochromatinization, that is condensation and inactivation of chromosomal material, creates a functional similarity to constitutive heterochromatin. The most widespread tissue-specific condensed euchromatin, sometimes called facultative heterochromatin, is found in sex chromosomes (for further details see Chapter 8). Furthermore, it is also true to say that heterochromatin will be constant in any given individual from cell to cell: inheritance is strictly mendelian.

The normal histological method of heterochromatin visualization is via the technique of C-banding, as shown in Figure 2.2.

True heterochromatin is not tissue specific. Accumulation of heterochromatin can be individual specific, but the position of accumulation does tend to be precise within the chromosomes. In most species heterochromatin is found flanking centromeres and telomeres.

When heterochromatin from centromere regions has been studied it has been found to contain tandemly repeated DNA, usually made up of simple repeats. This pattern seems to be consistent throughout all heterochromatin, but the nature of the repeated motif does vary. Such base repeat consistency within blocks of heterochromatin is

Figure 2.2 Human metaphase chromosomes which have been C-banded to show the distribution of constitutive heterochromatin.

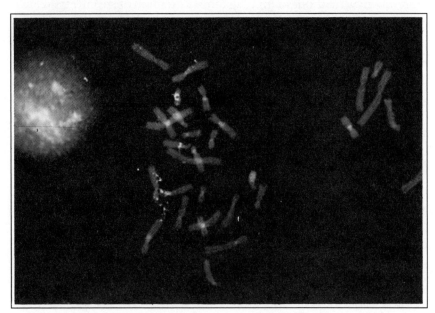

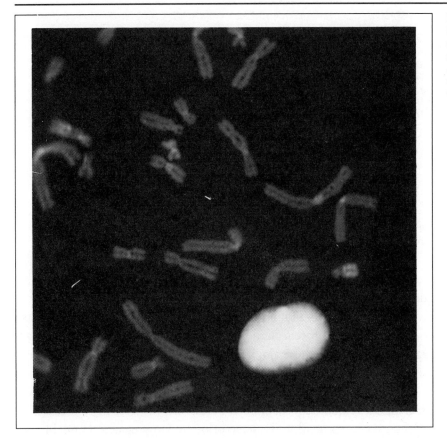

Figure 2.3 Distribution of heterochromatin repeat sequences on a human Y chromosome as shown by *in situ* hybridization. The distribution exactly matches that found by C-banding.

demonstrated in Figure 2.3. This shows the distribution of heterochromatin within a human Y chromosome, as found using *in situ* hybridization and corresponds exactly with that found using C-banding (for a description of C-banding, see Chapter 3).

A similar result has been found using restriction endonucleases of various sorts, depending only on there being no recognition sites present in the heterochromatin repeat motif. (See Chapter 3 for further details on restriction enzyme banding.)

Some repeat sequences have a significantly different DNA composition to the rest of the genome (for example being more GC or AT rich). Therefore they can be isolated as a separate band when whole genomic DNA is subjected to density-gradient centrifugation. These are known as satellite DNAs. The first to be isolated in such a manner was from the mouse. Since then they have been found widely throughout both the plant and animal kingdoms. The percentage of the total genome that is made up of heterochromatin can vary widely, as is shown in Table 2.1.

Table 2.1 Variation between species of genomic heterochromatin content as a percentage of the total (Mahan and Beck, 1986; John and Miklos, 1987).

Species	Heterochromatin (%)
Deer mouse (*Peromyscus crinitus*)	6
Barking deer (*Muntiacus muntjak*)	15
Human (*Homo sapiens*)	17
Vole (*Microtus agrestis*)	26
Mouse (*Mus musculus*)	27
Golden hamster (*Mesocricetus auratus*)	34
Drosophila novamexicana	34.4 (F), 28.4 (M)
Deer mouse (*Peromyscus eremicus*)	36
Drosophila americana americana	38.5 (F), 38.2 (M)
Drosophila americana texana	40.5 (F), 36.1 (M)
Drosophila virilis	48.3 (F), 49.2 (M)
Kangaroo rat (*Dipodomys ordii monoensis*)	58

The variation in heterochromatin content between closely related species can be clearly seen.

The central motif of satellite repeats can vary widely but tends to be quite short, such as GAAGAA/G in wheat or TAACCCC in felines.

Since we know that heterochromatin is made up of simple repeats, does not code for proteins and yet can make up a significant part of an organism's genome, the major questions surrounding it must be 'where is it located and what does it do?'.

General rules for the distribution of heterochromatin have been formulated, which suggest that C-bands are preferentially located at telomeres, centromeres and nucleolar organizing regions. Large blocks of telomeric heterochromatin are also supposedly associated with short chromosomes; the human Y chromosome is a good example of this. Non-homologous chromosomes have C-band material located at similar sites with respect to the centromere-band distance. *In situ* hybridization has been used to demonstrate some of these points.

Work using plant material has resulted in many suggestions regarding the mechanism of control of heterochromatin accumulation. It is, of course, essential that such control is exercised because even if the heterochromatin is non-coding, or junk, uncontrolled accumulation could result in severely impaired function if only because it would be expected that cell cycle time would be related to genome size. Although, as will be shown later, this does sometimes happen, the situation that is found is not always that simple.

Figure 2.4 tells an interesting tale in this respect. It can be seen that as the percentage heterochromatin content increases in species that have the same number of chromosomes, the cell cycle time does not alter in any consistent manner. This is exactly what we would expect from work on variation within a species. It is generally true that heterochromatin

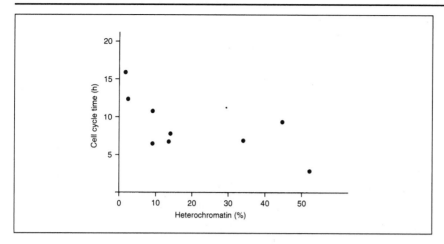

Figure 2.4 Cell cycle times versus heterochromatin content (%) for eight species of Compositae, all of which are $2n = 18$. It can be seen that there is no direct relationship between heterochromatin content and cell cycle times.

replicates late in the cell cycle, so replication must be coordinated in some way. Some species show large variations in heterochromatin content from individual to individual. Unless such variation also involves control of synchronous replication it would be expected that cell cycle times would vary by very large amounts. This would be especially true in cases where heterochromatin replicates last of all; without control, cell cycle times would be a direct measure of heterochromatin content. Control of heterochromatin is also important because the distribution of the chromosomes within the nucleus is now known to be fundamental to the control of transcription of the active genes. Thus, if accumulation of heterochromatin is random and uncontrolled, the genetic constitution of the individual would be severely compromised.

The question most frequently asked about heterochromatin is 'what possible function could it serve?'. Indeed, ever since it was first described in humans, correlation with some sort of phenotype has been sought, generally without success. Attempts have been made to correlate human heterochromatin with reproductive wastage, malignant disease or even criminal tendencies, and many have tried to measure heterochromatin variation at the population level. All such studies have been hampered by the method of measurement being totally subjective. To make meaningful measurements that will reflect what may well be a very small correlation, it is not enough to use nominal or ordinal data, scaled interval or ratio data must be used. This will hold true no matter what type of data are being measured. In the Nematoda *Ascaris* and *Parascaris*, heterochromatic chromosomes are eliminated wholesale during early embryonic cleavage. This together with the observation that *Drosophila* embryos have a complete complement of satellite DNA but that during the production of polytene chromosomes the heterochromatin is under-replicated such that satellite DNA is barely detectable in the late stages of polyteny, suggests that heterochromatin may be of little or no use to somatic cells but in some way of pivotal importance to the germ cell line. Certainly, the growth rate of larvae of the newt species *Triturus* is influenced by genome size, the variation in which is mostly dependent on the heterochromatin content of the cells.

The potential importance of heterochromatin in speciation can be overemphasized. Certainly, in the case of *Eulemur* species (a genus of lemurs) it has been found that, although there is considerable variability in the number of chromosomes, most species studied have a fairly uniform DNA content, with one exception however: *Eulemur coronatus*. Although *E. coronatus* ($2n=44$) has a far larger DNA content than *E. macaco* ($2n=46$) hybrids of these two species are viable because the additional DNA is almost exclusively heterochromatin.

It is not now generally regarded as likely that theories as to the nature and use of heterochromatin, such as the bodyguard hypothesis, reflect the importance of these sequences. The bodyguard hypothesis states that heterochromatin is in some way distributed about the nucleus so as to protect euchromatin from environmental damage. If this were true, of course, species with large heterochromatin contents would have much lower mutagenesis rates; experimentation tells us that this is not the case. While the accumulation of satellite DNA may be simply a biological phenomenon associated with heterochromatin, heterochromatin may be of a greater significance in evolution, being able to influence recombination rates when an organism's long generation time might require an additional mechanism to allow for rapid adaptation to a changing environment.

2.3   Telomeres

Telomeres are the end caps of chromosomes (from the Greek *telos*, end, and *meros*, part). They were postulated and named by H.J. Muller before even the significance of DNA was realized.

Telomeres are made up of both protein and DNA. It is the DNA sequence that is the essential component in the control of chromosome integrity. There are distinct similarities between those few telomere-binding proteins that have been looked at in any detail. The best known are those of the ciliate protozoa of the genera *Oxytricha* and *Euplotes*.

Oxytricha species have two proteins, one of 56 kDa and the other of 41 kDa. *Euplotes* species, on the other hand, have a single 51-kDa polypeptide. The DNA binding sites of the 51-kDa and 56-kDa proteins are very similar.

A protein very similar to these ciliate telomere-capping proteins is found associated with *Xenopus* telomeres. This level of protein conservation would imply that it may be a common feature of all eukaryote chromosomes.

When a chromosome breaks, losing the terminal DNA sequences, then the chromosome end is severely compromised in both form and function. One outcome of this is that unprotected broken ends can fuse, resulting in translocations, or ring chromosomes if both telomeres are lost from the same chromosome. Although this is important it is secondary to the fundamental function of telomeres, which is to stop the degradation of the chromosomes during replication.

The distribution of telomere sequences at inappropriate positions, such as centromeres, indicates that in the past chromosomes from various organisms may have fused, leaving the original telomeres high and dry.

2.3.1 The need for telomeres

If a chromosome is not formed as a ring, as in the case of plasmids, then it must have two ends. Each end must terminate in some sort of functional cap to prevent shortening of the chromosome at each round of cell division (section 2.3.2). This is the function of telomeres.

Although there are some notable exceptions, the sequence motif of telomeres, $(TTAGGG)_n$, is highly conserved across phylogenies. Evidence is available that shortening of chromosomes does indeed occur in some somatic tissues, the implication being that telomerase activity is tissue specific. Investigation of tumours has shown that telomere length is not necessarily predictable, sometimes increasing and sometimes decreasing, depending on the circumstances.

If telomeres were to shorten progressively, then cell death might well be the outcome. *Drosophila* species have a novel mechanism of overcoming this problem that involves retroelements, which make the presence of a conventional telomerase unnecessary. Retroelements are found clustered around the chromosome ends and are composed of large numbers of terminal repeats that are unlike the G-rich telomeric repeats normally expected. Evidence suggests that in *Drosophila* telomeres are replenished not by telomerase activity but by the occasional addition of several kilobases of these retroelements to the chromosome end.

2.3.2   Mode of action

DNA replication takes place in the 5′ to 3′ direction, with the result that one strand can be synthesized to the end while the other chromosome strand will have an unsynthesized section at the end upon which the DNA polymerase is primed and to which the bases necessary for complete replication cannot be added. This is shown in Figure 2.5.

The shortened chromosome that results would eventually start to lose genes, which could disable the organism or cause premature death. If the process were to continue uncontrolled it would endanger even the survival of the species because the germline chromosomes would also be shortened with every round of division. This is prevented by the presence of telomeres, which are formed as a result of the action of the enzyme telomerase (Figure 2.5).

It has been suggested that there is reduced telomerase activity in somatic cells, which could explain why telomeres found in sperm are longer than in somatic cells and why telomeres become shorter with age in somatic tissue and slightly longer with age in sperm. In humans the telomeres are approximately 10 kb long in sperm and 15 kb long in somatic cells. Progressive rounds of division in somatic cells should, therefore, result in diminishing telomeres. This has been found to occur in a number of cases. It has been shown that introduction of TTAGGG repeats can induce telomere formation in many different mammalian cell types. Such so-called subtelomeric repeats are quite common and highly variable in both number and composition. Although there may be additional long-term uses for subtelomeric repeats, they are not essential for chromosome function.

The presence of subtelomeric repeats can be highly variable. For example, approximately 21 additional subtelomeric bands that are not

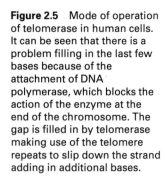

Figure 2.5 Mode of operation of telomerase in human cells. It can be seen that there is a problem filling in the last few bases because of the attachment of DNA polymerase, which blocks the action of the enzyme at the end of the chromosome. The gap is filled in by telomerase making use of the telomere repeats to slip down the strand adding in additional bases.

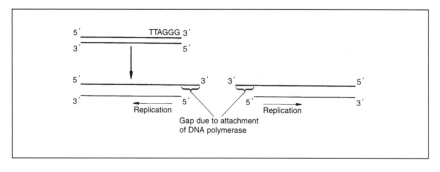

present in humans are found in chimpanzee karyotypes. This is despite the fact that there is 98% homology between chimpanzee and humans. This intriguing finding seems to be a result of the presence in chimpanzees of satellite DNA constructed of 32 bp which are AT rich. Although such DNA was also present in gorillas, it was not found in either orang-utans or humans. In the fission yeast *Saccharomyces cerevisiae* there are a number of mutations that directly affect telomere integrity. One such is called *est1* (Ever Shorter Telomeres), which results in progressive loss of telomeres and consequently aneuploidy and senescence.

In contrast, in the mouse (*Mus musculus*), there is no significant difference between the telomere lengths in somatic or germline tissue and there does not appear to be any progressive shortening of the telomeres with increasing age.

Telomerase from the protozoan *Tetrahymena* has been experimentally altered; as a result the base sequence of the telomeres also changed, resulting in phenotypes similar to *est1* in yeast. This form of yeast lacks the ability to synthesize terminal $G_{1-3}T$ repeats. From this it can be assumed that it is not conservation of the telomere sequence by conservation of ancillary proteins, but conservation as a necessity to maintain function.

Telomeres can be stained selectively by a method usually referred to as T-banding, which involves a series of stringent salt washes. On light microscopy T-banding reveals dark telomeres against a light background. However, electron microscope studies provide further structural information, revealing telomeres to be partially uncoiled DNA with loops of approximately 89 nm diameter.

2.3.3 Nature of telomere repeats

Repeat sequences that seem to delineate the ends of chromosomes have been found in telomeres. These repeats have the general sequence $(TTAGGG)_n$ (Table 2.2), although there are some startling exceptions to this, such as the hypotrichous ciliate *Stylonychia lemnae*, in which the repeat sequences follow the general plan $C_4A_4C_4A_4C_2$ and are not confined to the terminal parts of chromosomes.

The mechanism of replication of telomeres is pivotal in understanding maintenance of chromosome integrity. Vertebrate chromosomes are sometimes found in which telomere repeats are not exclusively confined to the ends of chromosomes. In these chromosomes it is suggested that the repeats are localized around the centromere, possibly

Table 2.2 The sequence repeat of the telomeres of various plant and animal

Telomeric repeat	Organism
TTAGGG	*Homo sapiens* (man)
TTAGGG	*Physarium* (slime mould)
TTAGGG	*Didymium* (slime mould)
TTAGGG	*Neurospora* (filamentous fungi)
TTAGGG	*Trypanosoma* (protozoan)
TTAGGG	*Crithidia* (protozoan)
GGGGTT	*Tetrahymena* (protozoan)
GGGGTT	*Glaucoma* (protozoan)
GGG(GT) TT	*Paramecium* (protozoan)
GGGGTTTT	*Oxytricha* (protozoan)
GGGGTTTT	*Stylonichia* (protozoan)
GGGGTTTT	*Euplotes* (protozoan)
AGGGTT(TC)	*Plasmodium* (protozoan)
TTTAGGG	*Arabidopsis* (plant)
TTTTAGGG	*Chlamydomonas* (alga)
$(A)G_{2-5}TTAC$	*Saccharomyces pombe* (yeast)
$G_{1-3}T$	*Saccharomyces cerevisiae* (yeast)
$G_{1-8}A$	*Dictyostelium* (slime mould)

representing ancestral DNA left when two chromosomes fused to form a single entity. When end-to-end associations of chromosomes are seen in cells it is nearly always in tumour cell lines or in disease characterized by DNA repair defects, such as Bloom's syndrome. The chromosome associations are thought to be due to telomere reduction, as a result of errors of DNA replication and repair. It has proved possible to demonstrate that when cell lines are set up *in vitro* telomere shortening takes place with each round of cell division. Once cells are transformed so that they are immortal then not only does telomere shortening cease, but telomerase activity can be detected. The telomeres, however, do not increase in length; it is as if telomerase activity is under very close control, merely stabilizing the telomere at the new length. Using probes to the telomere repeats of *Hordeum vulgare* and *Secale cereale* of the form $CCCTAAA_6$ and $TTTAGGG_6$, it has been shown that at least in these plant species the telomere repeats are confined to the chromosome ends.

It is now thought that the DNA at the extreme ends of telomeres is in the form not of a double helix but a single strand that is G rich and which protrudes beyond the complementary C-rich strand to form the hairpin structure necessary for telomere integrity. The highly conserved nature of the telomere sequences was first discovered when it was shown that the telomeric sequences of *Tetrahymena thermophila* (the first organism in which the telomere was sequenced) would cross-hybridize with human telomeres. Table 2.2 shows the telomere sequences for a variety of organisms. Conservation of telomere sequences has been investigated in over a hundred different vertebrates.

2.3.4 Telomeres and chromosome healing

The enzyme responsible for replication of telomeres is telomerase. This enzyme would seem to have a tissue-specific activity level. It has been suspected for some time that broken chromosomes somehow reform telomeres, but the way in which this occurs is unknown. This assumption is based on a simple observation that chromosomes with terminal deletions do not always get smaller with each round of cell division. Evidence that telomerase heals the broken ends of chromosomes is available, although it is not certain that this is a universal feature of telomerases. Healing of reduced chromosomes by telomerase has been clearly demonstrated in *Ascaris lumbricoides*. This process is a natural result of the chromatin loss that occurs in presomatic cells of *Ascaris* embryos.

Chromosome healing has been demonstrated in yeast chromosomes, which, unlike human chromosomes, can maintain cellular integrity with a greater mutational load. Also in yeast, it has been shown that chromosomes with one missing telomere can survive, but that chromosome with two damaged telomeres are quickly eliminated. Careful investigation has revealed a further use of telomeres in yeast: to allow the cell to recognize damaged DNA. A cut end without a telomere would be recognized as abnormal. This use of telomeres as points of recognition for intact chromosomes may be quite widespread, which would account for the process of chromosome healing observed in other organisms.

In humans it would be reasonable to assume that relatively small deletions of terminal material would result in haploid insufficiency that could prove detrimental. Candidates for such conditions are Wolf–Hirschhorn syndrome, in which the short arm of chromosome 4 is deleted, and Miller–Dieker syndrome, which is associated with deletion of the short arm of chromosome 17.

It has been suggested that in tumour cells chromosome breaks and cryptic translocations can be stabilized by the capture of telomere repeats from another chromosome. This could be of considerable importance in the aetiology of chromosome 6 involvement in melanomas, although how much is causal and how much is secondary remains to be determined.

Telomere length has been studied in intracranial tumours: of 60 samples from different patients, using peripheral blood for comparison, telomeres were longer than normal in 41%, shorter than usual in 21.7% and approximately the normal length in 36.7%. It would appear that telomerase activity was disrupted, being sometimes overexpressed, sometimes underexpressed and sometimes unchanged. A more clear-cut example is that of human ovarian cancer, in which activation of telomerase has been found. This will, as described earlier, prevent cell senescence and death by chromosome loss.

It would seem reasonable to suppose that in the case of most tumours of normally non-proliferative tissue, activation of telomerase is required to maintain cell integrity. In humans it has long been known that telomeres in germline cells are longer than those in somatic cells. This implies reduced telomerase activity in somatic cells, with a progressive loss of chromatin, and this has also been thought of as a method of programming cellular senescence. It has been noted that in some cell lines double-strand breakage results in arrest of cell division; this is termed checkpoint control. It could therefore be imagined that loss of telomere activity, as a result of progressive loss of repeat sequences, could also cause the cessation of cell division. Following on from this it has been suggested that the precocious immunosenescence of Down's syndrome patients is a direct result of excessive telomere loss. Although when Muller first postulated the need for telomeres to protect chromosome ends he was working with *Drosophila*, these organisms do not share a common telomere arrangement with the majority of eukaryotes.

It has been suggested that, since there does not seem to be a telomere repeat in *Drosophila*, another mechanism must maintain chromosome integrity. Over a period of 40 generations cultured flies were found to lose an average of 76 bp per fly generation. This could not, of course, go on forever without complete chromosome loss, if only because this is a rate of loss similar to DNA loss rates for broken chromosomes. One possible explanation is that a transposon is adding large blocks of sacrificial DNA to the chromosome ends at intermittent periods. Because the amount which can be added at one step is very large it becomes apparent that it may not be possible to see the event over 40 generations; possibly the addition occurs only when a very rare and specific trigger point is reached.

The importance of telomere structure and function is only now becoming fully appreciated, but one of the most important aspects of telomere genetics has been to explain the manner in which trypanosomes, *Neisseria gonorrhoea* and *Borrelia* can evade the humoral immune response. The trypanosomes have a variable surface glycoprotein, the expression of which is controlled by the physical position of the gene

Box 2.1   How the trypanosome swaps coats

The protozoan eukaryotes of the genus *Trypanosoma* are flagellate parasites. They are surprisingly common as vertebrate parasites, although normally the host is asymptomatic. It is only in humans and their domestic animals that these organisms appear to be pathogenic.

In Africa *T. gambiense* and *T. rhodesiense*, which are transmitted by the biting tsetse fly (*Glossina* sp.), are the cause of sleeping sickness. In South America *T. cruzi*, which is transmitted by a large hemipteran bug of the genus *Rhodnius*, causes Chagas' disease. Some species of *Trypanosoma* seem able to avoid the humoral immune response of the host. The method by which this is achieved is intimately associated with the telomeres of the parasite. *T. Brucei* has approximately 20 large chromosomes, arranged more or less in pairs, although their size varies considerably between homologues, and approximately 100 mini-chromosomes.

Each trypanosome carries about 1000 copies of genes that can code for the variable surface glycoprotein (VSG), giving the trypanosome a unique antigenic coat. Switching from one antigenic VSG to another allows the trypanosome to stay ahead of the host immune response. The only expressed VSG is found associated with the telomere of one of the chromosomes. To change from one to another the VSG gene has to be translocated to a control region close to a telomere. Such events seem to occur at the rate of about 10^{-2} to 10^{-6} per cell division; the infection waxes and wanes as the current VSG is recognized by the immune system and the mutation to another VSG results in a different population growing in the host as the old one is cleared (Figure 2.6).

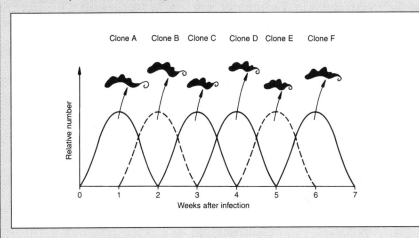

Figure 2.6 The course of trypanosome infections. Usually at the time of infection a small number of parasites are introduced to the host. These proliferate (as clone A) and there will be VSG gene translocations in a few individuals, so that by the time the host immunological system is responding to the infection a new clone (B) is already proliferating. Clone A is wiped out as clone B is increasing in numbers. This proceeds through several clones with a continual cycle of relapsing as the numbers of active trypanosomes in circulation increases and decreases with time.

on the chromosome with respect to the telomere. This is shown in Box 2.1.

2.4   Centromeres and kinetochores

The primary constriction of chromosomes is the centromere. It is characterized by particular repeat sequences of DNA and also by specific associated proteins. This is the last point of separation of sister chromatids during cell division. Movement of chromosomes is controlled by complex interactions between the kinetochore, a composite and robust structure that is associated with the centromere, and microtubules constructed of tubulin and actin. Although most centromeres and kinetochores are single structures, some species have so-called diffuse kinetochores which operate in the same way but throughout the entire length of the chromosome.

2.4.1   Centromeres

This enigmatic part of the chromosome is certainly of primary importance to the normal functioning of the chromosome within the cell, but it is also difficult to be precise about what makes a centromere. One of the most important aspects of the centromere is that it is the site of the kinetochore: this will be dealt with separately.

Satellite DNA of various sorts characterizes centromeres. In mammals these are, for the most part, alphoid satellite repeats of approximately 170 bp which characterize much heterochromatin. This DNA is present in every chromosome centromere. The range of repeat sizes varies in humans considerably, from 5 bp to the 170 bp of the alphoid repeat. Just as new candidate repeats are added to centromere structure, so sequences previously suggested to be centromeric are being shown not to be present in centromeres.

Interestingly, the repeat sequences which have been found are often highly conserved between organisms. It is also true that they need not necessarily be present on every chromosome of that organism. Consequently, we can say that beta-satellites of a 68 bp repeat have been found, as have other satellites of 5 bp or 48 bp repeats. In addition, a GGAAT repeat seems to be extremely common and an AT repeat also seems to appear frequently at centromeres.

Investigations involving the fission yeast *Schizosaccharomyces pombe* have shown that the centromere sequence Cen3 is slightly different to Cen1 and Cen2, which are a little shorter and with a slightly different repeat motif. In Cen3 there is a repetitive region of approximately 110 kb. This is made up of a central unit of 15–20 kb with flanking regions of 30 kb and 60 kb.

Although without doubt it is the DNA sequence that characterizes centromeres, it is not DNA alone which makes them what they are. It is possible that the sequence is pivotal in the construction of a self-assembly centromere in the form of a chromatin folding code.

An important part of any centromere is the associated protein. These are called CENPs. As a complete structure the presence of hetero-chromatin satellite sequences reflects something of the inert nature of this structure whereas the 17-bp CENP-B protein-binding motif (5'-CTTCGTTGGAAACGGGA-3') reflects the structural complexity. The suggestion that these repeats may occur throughout the alphoid satellite DNA and may form potentially functional CENP-B binding sites elsewhere along the chromosome is of particular interest since mobile centromeres and multiple centromeres are not unknown. We can learn much about the activation and inactivation of centromeres from abnormal chromosomal material.

It is generally thought that as an experimental organism *Homo sapiens* leaves a lot to be desired; this, however, is not always the case. The large-scale screening of patients for chromosomal problems, perhaps associated with an unfavourable obstetric history, has enabled genetic-ists to build up a picture of chromosome behaviour from repeated encounters with what were thought to be rare events. Thus it is that isochromosomes – especially human X chromosomes – will have two centromeres, but only one will be a functioning unit, the other virtually disappearing. The centromere is not necessarily therefore a permanent fixture of the chromosome but a dynamic structure. Figure 2.7 shows just such a human isochromosome X. To explain what has happened and show the position of the active and inactive centromeres, Figure 2.8

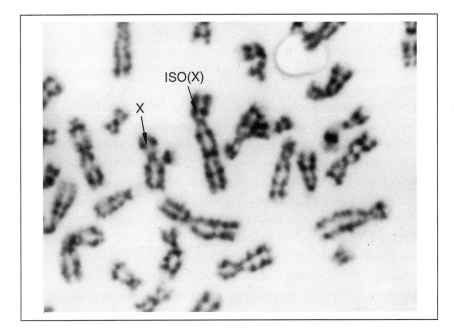

Figure 2.7 Human female metaphase with one normal X chromosome and one isochromosome X. Only one centromere is visible on the isochromosome. This individual is effectively trisomic for the X chromosome. Photograph courtesy of T. Spencer.

Figure 2.8 (a) The normal and (b) the isochromosome X as shown in Figure 2.7. The chromosome, as only one active centromere, which can be clearly seen, and an inactive centromere, the position of which is marked. The labelling is for ease of understanding the construction of the isochromosome.

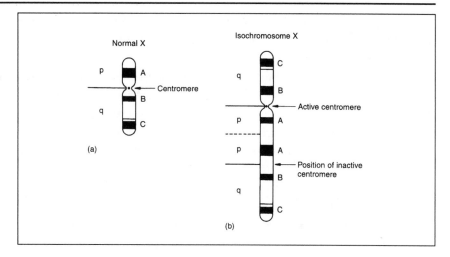

is a diagram of the normal and isochromosome X. It can be seen that the isochromosome is constructed by end-to-end replication from the short arm. The labels are simply to show the way that the chromosome is constructed. Only one centromere is active, as would be expected if the chromosome is to be stably inherited through normal cell division. If both centromeres were active it would be possible to lose the chromosome completely during mitosis.

The mouse cell line L929 has a marker chromosome with eight prominent centromeres and two minor ones, but only one will function as a centromere for the attachment point of the kinetochore. When cells of this type are grown in the presence of DNA intercalating agents such as H33258 the condensation of the satellite DNA of the centromere, shown to be 1.691 g cm^{-3} by centrifugation, is severely disrupted and clearly demonstrates that only one centromere is truly functional.

The first centromere to be studied in great detail was that of the human Y chromosome. This has yielded information which is of particular interest as it indicates the sort of repeat numbers which are required to generate a stable centromere which can cope with the complexities of cell division. It seems that 300 kb of material from the chromosome short arm, associated with about 200 kb of alphoid repeats, will result in a functional centromere. Deletions resulting in loss of alphoid repeats only will render a centromere inactive. On this basis then we can say that, although not necessarily the only essential component, alphoid repeats are pivotal to centromere function.

The size of the centromeric heterochromatin is polymorphic, but the variation in size of these units should not be taken as an indication of a change in function. Just as multiple centromeres are adaptable, so too can a range of centromere sizes be accommodated within a cell.

The polymorphic nature of centromeres is of importance in maintaining chromosome integrity. There does not appear to be a smallest functional unit below which a centromere cannot function. Chromosome breaks resulting in apparently stable telocentric chromosomes are known, as are Robertsonian fusions, resulting in a single chromosome with a single centromere. Figure 2.9 shows a C-banded human metaphase that has undergone a fusion between chromosomes 14 and 21. These two chromosomes have fused at the centromere and produced a single functioning unit.

Maize supernumerary B chromosomes, although genetically inert, must have all the requisite parts of a normal chromosome to be stably inherited at cell division. One such part is, of course, the centromere. Not only do B chromosome centromeres share many common repeat

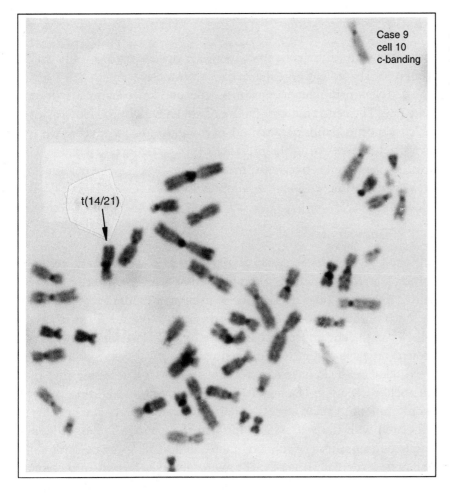

Case 9
cell 10
c-banding

t(14/21)

Figure 2.9 C-banded human chromosomes showing a centromeric fusion between chromosomes 14 and 21 resulting in a single functioning centromere. Photograph courtesy of S.C. Rooney.

Figure 2.10 Relationship between centromere and kinetochore.

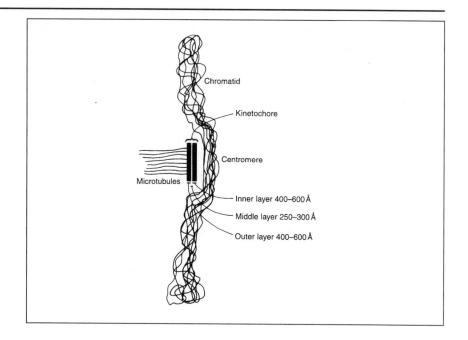

sequences with A chromosomes but also unusual repeat motifs which seem to be shared with the maize neocentromere.

The relationship between centromeres and kinetochores is an intimate one. The centromere is the site of attachment of the kinetochore, which is the most important feature of the chromosome involved in cell division. In other words, the presence of a kinetochore, or kinetochore proteins, is how an active centromere may be defined. The relationship between centromere and kinetochore is shown in Figure 2.10.

2.4.2   Kinetochores

Although the centromere can be regarded as the site of the primary chromosomal constriction and also the last point of sister chromatid pairing during anaphase onset, one of its most important features is the presence of the kinetochore. This is the anchor point for microtubule proteins at the chromosome end of the spindle. Two kinetochores are present on each chromosome; eventually one will face each pole. To avoid teleology it is necessary to explain how it is possible that two kinetochores end up facing in different directions. During mitosis microtubules grow and shrink, apparently in a random manner. This is associated with the self-assembling nature of microtubules. Above a certain concentration of tubulin, microtubules will spontaneously self-assemble. Eventually one of these microtubules will come into contact with the kinetochore and attach to it. As a result of a balance between movements towards and away from the poles, an attachment to the other kinetochore will orientate the chromosome correctly at the equator. This is important for accurate nuclear division. When kinetochores become malorientated the result is abnormal disjunction. Should this occur during meiosis unbalanced gametes can be produced – either monosomic or disomic (for more details on chromosome division and the spindle apparatus, see Chapter 4).

Kinetochores seem to be modelled on a general plan related to presence of centromere-associated proteins:

CENP-A   17 kDa
CENP-B   80 kDa
CENP-C   140 kDa
CENP-D   50 kDa.

It has been shown that in stable dicentric chromosomes CENP-B is present at both active and inactive centromeres, although monocentric chromosomes vary widely in their CENP-B content. CENP-C, on the other hand, has not been detected at inactive centromeres. This observation has led to the idea that CENP-C is vital to the normal functioning of the kinetochore, but is not necessarily a component of the generalized centromere. More recently, the distribution of CENP-C has been localized to the kinetochore plate. This would imply that, although other CENP proteins are essential for normal kinetochore structure and function, CENP-C is a functional part of the kinetochore. In studies of stably inherited dicentric chromosomes it has been found that, while both chromosomes contain CENP-B, only the active one contains CENP-C. Using antibodies to CENP-C proteins it has proved possible to disrupt mitosis in HeLa cell lines.

Almost all (99%) CENP-B is found outside the kinetochore and generally arranged in the centromere. CENP-B has two regions that are acidic because of their high glutamine and aspartamine content. These could alter the chromatin conformation by reacting with other proteins present at the centromere. The kinetochore is specifically the binding site for the microtubules which pull the sister chromatids apart during anaphase of cell division. Other chromosomal binding proteins are responsible for holding the centromeres together while the microtubules contract and separation starts.

There are some groups whose chromosomes do not have single clearly defined centromeres and kinetochores. Some insects as well as monocotyledons fall into this category. Their chromosome lacks a primary constriction and so a diffuse kinetochore is constructed down the entire length of the chromosome. The most obvious result of this is that, as nuclear division proceeds, chromosomes do not bend at the middle with the kinetochore leading the way polewards. The chromosomes remain parallel as they progress through anaphase. One advantage of this type of kinetochore organization is that damage from ionizing radiations which break chromosomes does not result in expulsion from the nucleus of an acentric fragment. All parts of broken chromosomes still proceed to the nuclear poles.

There is a group of mammals that have undergone several translocations which have resulted in compound kinetochores of remarkable size, although not as extensive as the diffuse kinetochores. The Chinese muntjac (*Muntiacus muntjak reevesi*) has $2n=46$. All of these chromosomes are small and telocentric. This subspecies is almost indistinguishable from the Indian muntjac (*Muntiacus muntjak vaginalis*), with which it can form hybrids in captivity. The Indian muntjak, however, has $2n=6$ in the female and $2n=7$ in the male. This surprising state of affairs seems to be largely due to centric and tandem fusions that have caused the reduction in chromosome number without disrupting the genetic integrity of the species. By having a translocation between an autosome and X chromosome the male ends up with an additional chromosome. By the same process the X chromosome has a massive centromere and associated kinetochore which constitute more than one-third of the length of the chromosome.

The remarkable resilience of metaphase centromeres to disruption by proteolytic enzymes, hypotonic solutions, chelating agents and strong alkaline solutions seems to be associated with the kinetochore. CREST serum, which contains antikinetochore antibodies, still reacts with the centromere even after all associated histone and non-histone protein has been removed from the chromosomal DNA scaffold.

Using a modified silver staining technique called Cd-banding it is possible to stain kinetochores, but the most reliable method is via the use of CREST serum. CREST serum originates from patients who have the mild autoimmune disease scleroderma. The acronym CREST comes from the major symptoms of the disease:

- calcinosis
- Raynaud's phenomenon
- (o)esophageal dysmotility
- sclerodactyly
- telangiectasia.

The antigens recognized by CREST serum are very susceptible to damage by methanol/acetic acid fixative. For this reason, when using

CREST serum the fixative of choice is methanol, acetone or formalde-hyde alone. Once used CREST is visualized by immunological methods, such as fluorescein isothiocyanate (FITC)-labelled IgG, which can be seen when fluoresced by UV light.

This technique is most often used when studying dicentric chromo-somes, whose stability is related to having only one active kinetochore. Cells from older humans seem to lose Cd-bands. If there is no kinetochore the chromosome becomes unstable and so reflects the occasional somatic cell chromosome loss in elderly patients.

2.5 Nucleolar organizing regions

Nucleolar organizing regions, usually referred to simply as NORs, are exceptional among transcribed DNA as they lead directly to a visible structure in the nucleus: the nucleolus. It is this structure that is the site

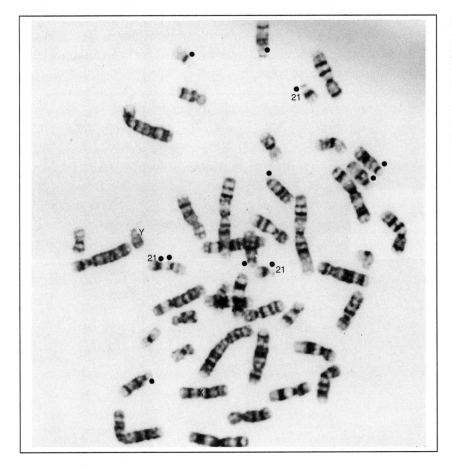

Figure 2.11 The distribution of NORs in the human karyotype. The chromosomes which have NORs associated with them are marked with a spot. Normally there would be 10 NOR-carrying chromosomes in a human diploid cell. In this case there are 11. This is because the karyotype is 47, XY, + 21: Down's syndrome. The three chromosome 21s are marked.

of ribosome formation. Ribosomal RNA is transcribed and processed in the nucleolus. If carefully prepared it is possible to demonstrate growing rRNA chains along NOR DNA by electron microscopy. The normal method of visualizing NORs is by the use of silver staining techniques, although in certain species, humans for example, NORs are distributed in such a way that their presence is known, even if they are not always visible. In the human karyotype NORs are to be found associated with satellited chromosomes; these are chromosomes 13, 14, 15, 21 and 22, as shown in Figure 2.11.

This association with satellites is not always the case, however. Figure 2.12 shows the distribution of NORs in *Pisum sativum* (pea); in this plant NORs can be seen to have a quite different arrangement.

That NORs are essential for normal development has been clearly demonstrated in *Drosophila*. The normal state of a complete set of NORs can be altered in the mutant phenotype 'bobbed' of *Drosophila*. The product of rDNA is rRNA of three broad types: 5S, 18S and 28S. These are cleaved and assembled with nucleoproteins to form ribosomes.

After mitosis nucleoli are quickly reformed by association of peri-nucleolar bodies with active NORs. It can therefore be said that the nucleolus contains some NORs. The nucleolus generally 'disappears' in late prophase and reappears in telophase (Chapter 4).

Figure 2.12 NORs in *Pisum sativum*. Photograph courtesy of Dr K. Hall.

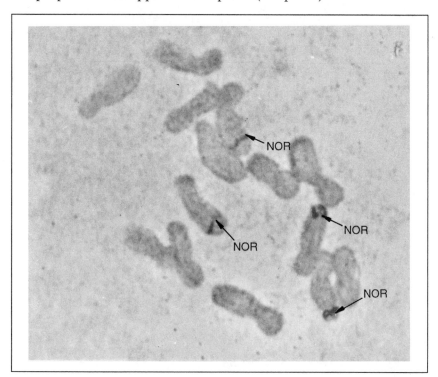

There does seem to be a certain amount of variation in the sequences of NORs, but with a quite high level of conservation between species. It is also apparent from cloned cells and twin studies that NOR sequences are stable throughout mitosis, although meiosis seems to generate new lengths and sequences quite readily, most likely as a result of unequal homologous exchange.

All known ribosomes contain both variable and invariable sequences. Twelve invariable sequences are found dispersed between the variable regions. The variable sections are almost species specific whereas the invariable sections are highly conserved across phylogenies. Such is the variation found that it has been suggested that no two ribosomes have the same sequence.

In human chromosome 21 it has been found that gene orientation is telomere to centromere, so that even with the large polymorphisms found in satellite stalk length the crossovers would be undetectable cytogenetically. Detailed descriptions of chromosome 21 has allowed us to be highly specific about the positioning of rDNA and associated satellites and satellite repeats, as shown in Figure 2.13.

There are two methods of visualizing NORs: silver staining with silver nitrate to highlight active NORs; and strong acid treatment, which produces N-banding, staining all NORs, active or not. In human cells it is almost impossible to stain all the NORs in a single cell using N-banding. There can be considerable polymorphic variation in the size of NORs of these chromosomes, the polymorphisms being relatively stable and therefore inherited in a mendelian fashion.

In some species N-banding can be a basis for the production of karyotypes as the NORs are scattered widely among the chromosomes. Such a case is found in *Hordeum vulgare* (barley). Figure 2.14 shows the distribution of NORs *in situ* hybridization rather than N-banding.

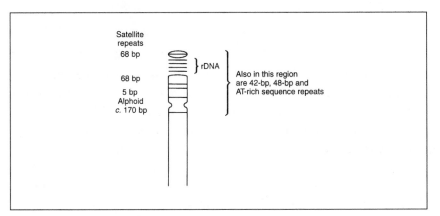

Figure 2.13 Structure of the satellite region of chromosome 21.

Figure 2.14   Barley NORs.

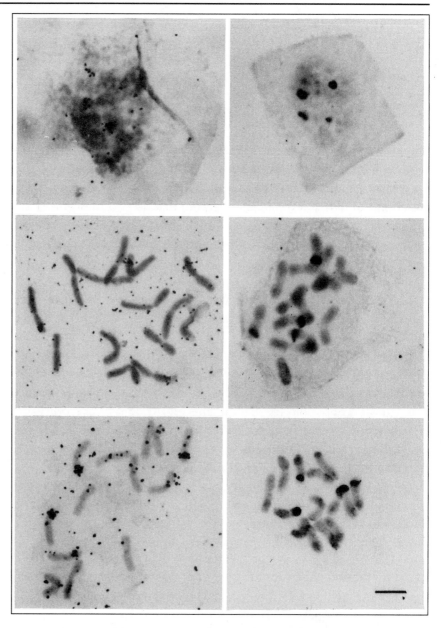

It is also possible to use N-banding to stain kinetochores, although this is not as efficient as the staining of NORs. It has been shown using chromosomes of *Xenopus* that N-banding removes all but 6% of the protein from chromosomes. Exactly how N-banding works is hard to imagine unless the remaining protein is at the site of NORs.

There does not seem to be any particular consistency in the numbers of NORs between species; thus, lower primates may have them on a single homologous pair of chromosomes, while orang-utan and mouse both have eight pairs of chromosomes with NORs. There certainly seems to be satellite associations between NORs in cells at metaphase, although whether this reflects some detail of chromosome disposition during interphase or an association artefact created during cell preparation is unknown.

It is the satellited chromosomes, those bearing NORs, that are involved in Robertsonian fusions, resulting in the loss of NORs, but this is not the only location of NORs. Mouse chromosomes frequently carry NORs near the centromere, but sometimes terminally. Both guinea pig and gorilla have telomeric NORs; the mechanism of distribution and control of NORs such as these is still unknown.

2.6 Lampbrush and polytene chromosomes

2.6.1 Lampbrush chromosomes

Lampbrush chromosomes are so named because of their unusual morphology. This is shown schematically in Figure 2.15.

When viewed as a whole, the large number of loops look like bristles on a brush. Their discovery by Flemming in 1882 explains the name 'lampbrush' as well; although lampbrushes are almost unheard of now, the modern equivalent is the bottle brush, and this is more or less the appearance of these chromosomes. The first working maps of lampbrush chromosomes were diagrams of the morphological features, such as size, centromeres and any other features that could be used to identify individual chromosomes and any rearrangements that might be found. These unusual chromosomes are a special, though not unique, feature of the oocytes of some amphibians. Lampbrush chromosomes

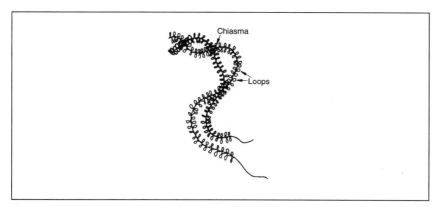

Figure 2.15 The appearance of amphibian lampbrush chromosomes. Two homologues are held together by chiasmata and fused centromeres. Although there is a similarity of appearance these are not metaphase chromosomes.

have not been described, for example, in caecilians. The chromosomes become elongated, up to 800 μm in length, during diplotene of the first meiotic prophase, a period in cell division which can last several months. There is a reason for this which is associated with cellular activity. Before fertilization material produced in the egg will direct early division and the fate of early cells. In addition, the embryo will not start production of ribosomal RNA until gastrulation has begun, and even then not in adequate amounts until hatching; as a consequence, any ribosomal RNA that is required by the organism must be provided at an early stage. The production of ribosomes is undertaken in a slightly different way in that circular ribosome genes are replicated so that complete production of 18S and 28S RNA can proceed quickly and efficiently.

It is informational RNA, protein-coding material that will use the additional ribosomes, that is transcribed from lampbrush loops, although apparently non-coding DNA also being transcribed from lampbrush loops has been demonstrated. After crossing over has occurred the two homologous chromosomes are held loosely together by chiasmata and fused centromeres. This gives lampbrush chromosomes the loose appearance of metaphase chromosomes with which they should not be confused. Loops of DNA from each single strand are unwound and thrown out; loops are therefore part of the structure and not imposed upon it (Figure 2.16). They are also in identical pairs, each pair being formed from a chromomere at the base of the loop. Extending the chromosome by physical manipulation straightens out the loops. Each loop is formed from about 50 nm of DNA, so most of the DNA is packed into the chromomeres.

The lampbrush number 1 chromosome of *Triturus carnifex*, a newt, has between 350 and 500 lateral loops. The transcriptionally active DNA seems to be associated with approximately 20–30 of these loops, which can therefore be used as landmarks.

Figure 2.16 Continuity of structure of lampbrush chromosomes. Loops are not imposed on a structure but make up the structure itself.

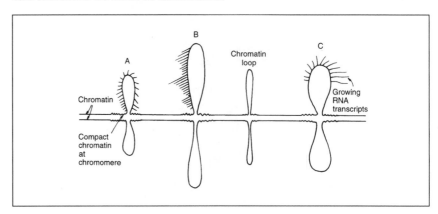

By using deductive reasoning it has been suggested that each loop, of which there are many hundreds, represents a single gene forming a single polypeptide. Transcription seems to start fairly close to the central chromosome axis and proceeds around the loop to the other side. As the RNA polymerase molecule moves along the DNA progressively longer RNA molecules are to be found associated with the loop, and soon after transcription starts proteins also seem to associate with the growing RNA molecules.

If it is assumed that each loop represents a single gene, then it is possible to say that the number of loops reflects the number of genes that are active at the time. This apparently small number of active genes reflects the requirements of the early embryos for a small number of polypeptides to direct early development. There does seem to be evidence that some loops have more than one gene on them, but in these cases each gene is separated by non-coding sequences. Similar non-transcribed sequences are found at either end of loops where only a single gene is present. By using monoclonal antibodies it has been possible to demonstrate that there are differences in the lampbrush chromosomes of different, closely related, species of newts; in this way the cytogenetics has been able to confirm the taxonomy of this group.

Besides amphibians, lampbrush chromosomes can be found in many different species, including humans, and in all these cases they are found in oocytes. In *Drosophila hydei* they are also present, but this time not in oocytes. In this particular case they are found in spermatocytes, the precursors of spermatozoa. There is also a morphological distinction which can be made between chromosomes. Only the Y chromosome becomes extensively looped and therefore of classical lampbrush appearance; other chromosomes of this species loop to only a limited extent. There is not only a difference between chromosomes, but also between loops on the Y chromosome itself. Five loops can be seen, each one distinct in position and morphology.

2.6.2 Polytene chromosomes

Polytene chromosomes are a special feature of dipteran flies, specifically of the genera *Drosophila* and *Chironomus*. This is not to say that they do not occur elsewhere; indeed, they are not unusual in plants, but they are rather more difficult to characterize. These very unusual structures, chromosomes clearly visible during interphase, are constructed by endoreduplication and are only found in certain very large cells, such as in the gut, excretory organs and salivary glands of dipterans, the last tissue being the one most frequently studied. This process is simply one of chromosome replication without nuclear division. In this way it is possible to produce tetraploid cells; human cells in culture are capable of undergoing endoreduplication by addition of colchicine to disrupt nuclear division. In this particular case the chromosomes usually do not remain attached to each other and endoreduplication occurs only once.

The induction of polytene chromosomes in plants has been demonstrated *in vivo* and also *in vitro*. In the latter case induction is an artefact of long-term tissue culture and heat shock of cell suspensions. This is presumably a reflection of the endoreduplication events caused by nuclear disruption. Plant polytene chromosomes tend to be poorly defined structurally when compared with dipteran polytene chromosomes; for this reason it is to this group that most studies have been confined. In *Drosophila* polytene chromosome endoreduplication continues for 10 complete rounds of chromosome replication; 2^{10} copies or approximately 1024 DNA strands. In *Chironomus* there may be 13 rounds of doubling.

Not all DNA is replicated in this manner; the rRNA genes, which are already present as multiple copies, are not. These genes would normally account for approximately 0.5% of the genome, but in polytene chromosomes they are only 0.1% of the genome. When replication is complete the separate strands of DNA are lined up exactly parallel. Figure 2.17 shows polytene chromosomes from *Drosophila melanogaster* larval salivary glands. A model of this is shown in Figure 2.18. Both telomeres and centromeres are supposedly under-replicated, but just as

Figure 2.17 A polytene chromosome.

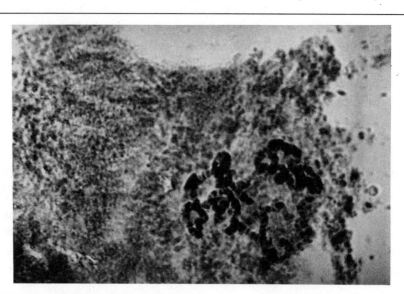

Puffs

Figure 2.18 A hypothetical structure of a polytene chromosome. It is suggested that there are two areas of under-replication: telomeres and centromeres. Ribosomal genes are replicated but not as much as the bulk of the euchromatin.

the construction of chromosome regions around rRNA genes is difficult to visualize it is also difficult to explain the reduction of multiple copies of DNA down to a single strand.

It is now thought that polytene chromosomes are special cases of lampbrush chromosomes, simply brought into intimate contact and precisely lined up. One of the most useful features of polytene chromosomes is that they can be handled in tissue culture so that changes in their structure and behaviour can be followed through time. Use of phase-contrast microscopy reveals that even unstained polytene chromosomes have a banded structure. This is in direct contrast to chromosomes, which have to be treated or digested with enzymes and then stained to reveal a banding pattern. Staining polytene chromosomes greatly enhances their visibility but destroys any activity. Any polytene banding pattern is constant within a tissue of a species but not necessarily between species or tissues. This should not be a surprise since, if there is a coordination between puffing and transcription, different tissues will have different requirements. It has been shown that in *Ceratitis capitata*, a small fly, the banding pattern of polytene chromosomes is tissue specific. By looking at two different cell types, salivary glands and bristle trichogen cells, the patterns were shown to be so different that the homologous chromosomes were not recognizable. Various strains of this species have different translocations and inversions which did allow the determination of the homologous chromosomes. Since the bands are condensed DNA and the interband is decondensed DNA, it can be assumed that the enormous difference in transcriptional requirements between these two cell types is reflected in the banding pattern.

Such is the level of detail available from the bands that, if there is an inversion in one of the two chromatids, this may well induce asynapsis and the extent of their rearrangement will then become obvious. The most significant question we can ask about polytene chromosomes must be 'what is the relationship between bands and genes?'. It would seem that each band corresponds to a single gene or transcriptional unit; this law may yet prove to have exceptions to it, but as a general rule it seems to hold quite well.

One phenomenon displayed by polytene chromosomes is that of puffing, the formation of decondensed DNA associated with bands. The bands are normally highly condensed; the interbands are barely so. Very large puffs are referred to as Balbiani rings. When puffed bands are looked at in more detail they often appear highly diffuse, but this is in part an illusion. Much of the appearance of a puff is the result of associated RNA and protein; actively transcribed sections of DNA make up the puff. The size of the puffed band is nearly always much larger than a functional gene would need to be, given that *Drosophila* does not have a set of polypeptides noticeably larger than any other organism. It has, however, been shown that the RNA product of Balbiani ring 2, which is on chromosome 4 of *Chironomus*, is many times larger than normal mRNA. This large molecule exists for only a very short period of time before being degraded, a process that is completed in the nucleus. Nucleases digest most of the RNA to free nucleotides, while the 3' end remains as the home of the required mRNA. It always seems to be the 3' end that is kept, so it can be said that usable RNA is transcribed last from the puffed region. It is processed RNA that is transported out of the nucleus.

Puffing can be induced in chromosomes by a variety of agents, such as hormones or heat shock. This should not be of particular surprise since any stimulus that alters a cell in some way must act through the medium of the genetic material. The insect moulting hormone ecdysone is an example of a hormone altering gene expression in some way. Addition of ecdysone to polytene chromosomes results in an altered but highly specific and repeatable puffing pattern.

2.7   Double minutes and homogeneously staining regions

These structures represent amplification of specific genetic material. They are found occasionally in mammals, in particular humans, and more frequently in cultured cell lines. They are generally associated with resistance to antimetabolite drugs. The classic example is amplification of the dihydrofolate reductase gene (DHFR) to overcome methotrexate (MTX) cancer chemotherapy. They are also associated with many human cancers, frequently appearing in tumour cell lines. A review of over 200 primary human tumours showed that 91% contained double minutes (DMs), 6.5% homogeneously staining regions (HSRs) and 2.5% both.

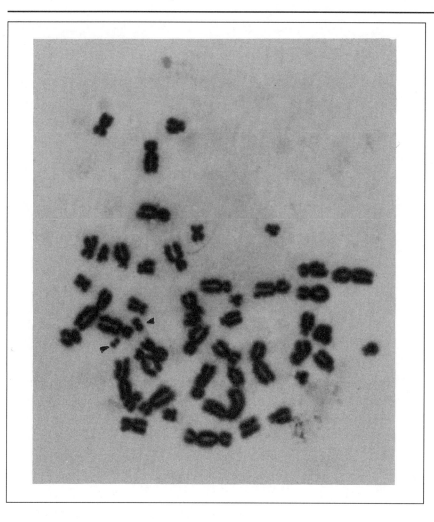

Figure 2.19 Cell showing DMs. Courtesy of A. Harvey.

2.7.1   Double minutes

DMs look like minute chromosomes (Figure 2.19). They contain approximately 1 000–2 000 kb of DNA each and are self-replicating with no discernible or active centromeres. They do not participate in regular division at mitosis, segregating in a random fashion and producing an unequal distribution in the daughter cells. They may also be lost through micronucleation, a process similar to that which affects univalents. Up to 20 have been found in MTX-resistant cell lines.

2.7.2   Homogeneously staining regions

HSRs are large regions of non-banding amplified material within a chromosome. The size of the HSR is related to the length of the

Figure 2.20 Methotrexate-resistant human leukaemia cell line showing HSRs in three marker chromosomes (large arrows). For comparison the normal chromosomes 6 and 9 are also arrowed. Reproduced with permission from Bertino *et al.* (1982) *Gene Amplification*, Cold Spring Harbor Laboratory.

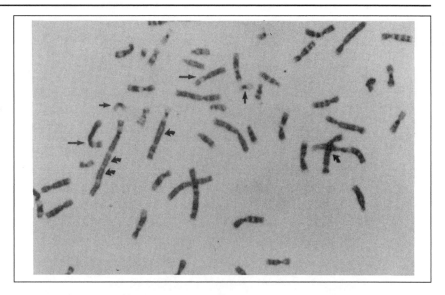

sequence amplified and the number of times it is multiplied. It is presumed that amplification occurs at the site of the resident gene responsible. For example, in drug resistance, the HSR found in mouse cell lines resistant to MTX is in the middle of chromosome 2, the site of the resident *dhfr* gene (which produces resistance to MTX). Multiple HSRs can occur in cells as a result of several amplification events or breakage and translocation of the original HSR.

These two forms of extra DNA are regarded as different expressions of the same genetic mechanism in that DMs are thought to be unstable precursors of HSRs. Evidence for this includes the observation that the longer drug-resistant cell lines are grown, the greater the substitution of DMs by HSRs. Also, it has been shown that DMs can be eliminated from mammalian tumour cell lines by the application of low concentrations of hydroxyurea, but this does not affect HSRs, i.e. the situation with DMs is reversible, but HSRs are stable additions to the cell.

The lack of DMs and HSRs in a drug-resistant or tumour cell line does not imply non-amplification, it could simply be that amplification has occurred, but at an undetectable level. The link between these amplification events and drug resistance is fairly well established, however the exact nature of the relationships with cancers is still unclear. At the moment it is not known whether amplification is related to either the initiation or progression of the disease, a totally unrelated consequence of the abnormal properties of the tumour cells or a mix of all three. Clearer definition relies upon the identification of the amplified

Box 2.2 Disproportionate replication

The central feature of the model is the probability that replication can be initiated at the same origin (or close by) more than once within any given synthesis phase of a cell cycle. In addition, it is proposed that the DNA replication and elongation process slows down or ceases at a site within the chromosome before joining with a replication fork from an adjacent origin of replication. The consequence of these additional rounds of replication would be the production of multiple DNA strands attached to the chromosome at one site, an 'onion skin' type effect (Figure 2.21).

These extra strands may be released in either a linear or circular form. In the absence of selection pressure, these may simply be lost from the cell with no effect. In the presence of selection, for example application of drugs, replication and ligation of these strands could generate DMs. End-to-end ligation followed by recombination with a chromosome could produce HSRs. How these events occur or are controlled is unknown, however they obviously involve the nuclear replication machinery.

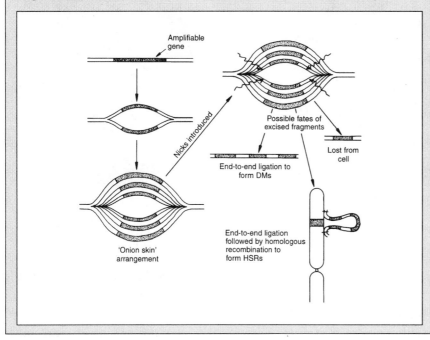

Figure 2.21 Saltatory or disproportionate replication. Replication can be initiated at the same origin several times during any one S phase, producing an 'onion skin' configuration. The potential fates of the 'extra' strands are shown.

sequences and their effects. Although the amplified regions represent relatively large areas of DNA for cloning, even if the limits of the gene and its sequence are defined it is difficult to identify the gene function if it is unrelated to other sequences in current gene sequence databases.

How do these structures originate? Schimke proposed a model based on disproportionate (saltatory) replication (Box 2.2).

2.8 B chromosomes

These are often called supernumerary chromosomes and are additional to the normal basic chromosome complement, occurring in over 1000

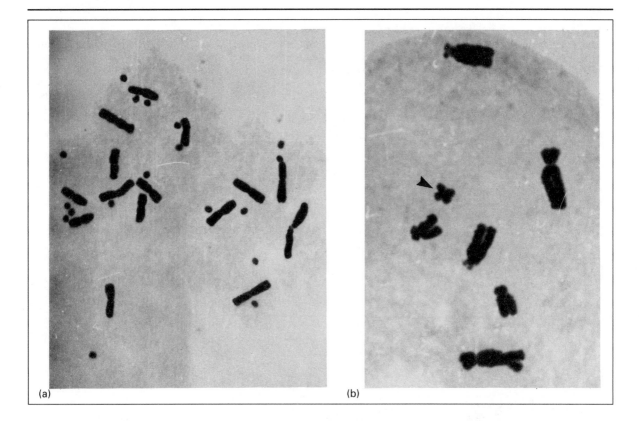

(a)   (b)

Figure 2.22 Examples of B chromosomes. Courtesy of Professor J.S. Parker. (a) *Allium schoenoprasum* cell with 17 Bs. (b) *Crepis capillaris* with a metacentric B (arrowed).

plant and 260 animal species, usually with a population-specific distribution (Figure 2.22). Individuals possessing B chromosomes are usually phenotypically indistinguishable from those without. The extra chromosomes are called Bs to distinguish them from the normal (A) complement. The main criterion for designation as a B chromosome is that they are dispensable and not essential for normal growth, development and reproduction. Most of the work to determine their form and function has been carried out on plants, as these are easier to manipulate experimentally.

The general characteristics of Bs are shown in Table 2.3. The origin of B chromosomes is thought to be from A trisomic autosome variants, with evolution acting on the univalent unpaired A chromosome via mechanisms similar to those operating on sex chromosomes (Chapter 8). Over time, this leads to unique B chromosomes. This is substantiated to a certain extent by the finding that some Bs share homology with certain regular A chromosomes, in particular the sex chromosomes. Early studies on DNA composition (base ratio, amount of repetitious DNA, etc.) showed no significant difference to the As. However, recent molecular studies have isolated unique B

Morphologically distinct from A chromosomes Smaller than As Contain large proportions of heterochromatin Do not pair with As during cell division Exhibit non-mendelian inheritance Do not carry major genes Influence primarily determined by the effects of their cumulative number	**Table 2.3** General characteristics of B chromosomes

sequences, indicating that, although there is homology, significant evolutionary changes have taken place since. Answers are starting to be obtained in this area as a result of the ability of some Bs to be separated by pulse-field gel electrophoresis and flow cytometry. This enables B chromosome-specific DNA libraries to be produced and analysed using molecular techniques.

B chromosomes do not generally contain major genes. Some 20 species (six insect and 14 plant) contain ribosomal DNA sequences, and there are a handful of other exceptions. A single B chromosome can produce male sterility in *Plantago coronopus* (Buck's horn plantain) and alteration of achene (seed/fruit) colour in *Haplopappus gracilis* (Compositae family), while the presence of five or more produce leaf striping in maize. The most curious example of B chromosomes has to be that of the primitive New Zealand frog, *Leiopelma hochstetteri*. The females of the species have an extra supernumerary chromosome that is distinct from the others, and this has been shown to be the sex determination system of this species (OW/OO).

The number of B chromosomes can vary between populations of the same species, within individuals of each population and even within cells of an individual. Variation within the cells of an individual can be due to mitotic instability caused by lagging at the metaphase plate and non-disjunction. However, in many instances, there must be some form of higher control as this individual cell specific variation is very defined. For example, Bs can be excluded from whole organs such as roots in *Aegilops speltoides* (wheat species) and *Haplopappus gracilis*. The B chromosomes of a Swiss population of *Crepis capillaris* undergo somatic non-disjunction in shoot meristems at the time of floral initiation, which results in an increase in the number of Bs in the flowers and therefore enhanced transmission through the germ line (Figure 2.23). The accumulation of Bs in the grasshopper *Myrmeleotettix maculatus* is effected through directed segregation in the oocytes.

As many as eight B chromosomes have been detected in wild plant species, although this number can be increased up to 34 as a result of

Figure 2.23 Behaviour of Bs at meiosis in *Crepis capillaris,* uneven distribution at telophase II. Photograph courtesy of Professor J.S. Parker.

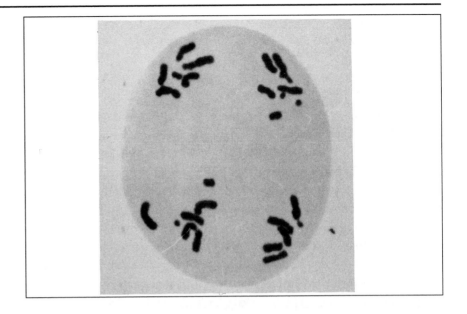

artificial manipulation. Among the vertebrates, some of the greatest supernumerary chromosome variation is shown by *Leiopelma hochstetteri*. Individuals have been found with up to 16 supernumerary chromosomes in a single cell. This variation in number in this particular instance is due to their instability during meiotic cell division via polarized non-disjunction.

Although individuals with B chromosomes are virtually indistinguishable from those without, B chromosomes cannot accumulate indefinitely within an organism because fitness is severely impaired when large numbers are present. Their effects are quantitative rather than qualitative, affecting total protein and RNA content, and producing a reduction in vigour and fertility. They have been shown to affect meiosis in the As, generally by increasing the synaptonemal complex length and the number of chiasmata; this effect may be beneficial for releasing genetic variation (although in some species the reverse is true in that the number of chiasma decrease, e.g. *Aegilops speltoides* and *Lolium perenne*).

It has recently been shown that the A chromosomes and the Bs themselves can influence the segregation of Bs at cell division, although this process and the genes involved are still far from being understood. If the numbers of Bs increase too much, the meiotic process starts to deteriorate. One peculiar phenomenon that occurs in some plants is a zig-zag effect on development, in which even numbers of Bs are more favourable than odd (the 'odds and evens effect'). It has recently been shown in rye that there is a maternal imprinting effect on B-chromosome inheritance. Crosses were made of 0B×2B, 2B×0B and 2B×2B plants. If the maternal parent had two Bs, transmission of Bs was increased, whereas if the maternal parent had no Bs there was a decreased transmission of Bs to the offspring. A similar type of imprinting mechanism may affect other species as different populations of the same species often exhibit different rates of B transmission. It has been suggested that this imprinting effect may be related to methylation or demethylation of control sequences on the A chromosomes (described in Chapter 5.).

It has been argued that Bs are nuclear parasites, maintained within a population only as a result of accumulation mechanisms. However, this cannot solely account for their widespread distribution. It has been shown that they can confer superior fitness on an individual or population under stressful conditions. For example, it has been shown that riverside populations of a member of the onion family (*Allium schoenoprasum*) with Bs survive drought and flooding conditions better than individuals without Bs. A specific stress or geographical pattern has not been found for many organisms containing B chromosomes. However, their effect on meiosis and alteration of chiasma patterns may well prove to be an important mechanism for potential adaption to changing circumstances via novel recombination events releasing new genetic variation or have evolutionary consequences.

2.9 Supernumerary segments

In some species, particularly grasshoppers (Orthoptera), polymorphisms exist within populations for extra segments on particular chromosomes. These are called supernumerary segments by analogy with supernumerary (B) chromosomes, with which they share several characteristics, i.e. they are often heterochromatic and dispensable, and individuals possessing them are indistinguishable from the normal population.

Although precise functions have not been assigned to these segments, they are by no means genetically inert. In most grasshoppers they increase the mean cell chiasma frequency and produce changes in the distribution of chiasmata within and between chromosomes. Those chromosomes possessing segments exhibit preferential chiasma formation in the regions furthest from the segment. This effect of redistribution of chiasmata (and, presumably, increased release of genetic variation) has been shown, in some cases, to have adaptive significance. Where several different types of segment have been found in the same cell, there is no evidence of interaction between them. However, the segments have been shown to interact with the nuclear genome, such that identical segments have different effects on chiasma frequency in different populations. Similar, again, to B chromosomes, the effect of an increasing number of segments has both an additive and 'odds and evens' effect.

The other main example of possession of supernumerary segments apart from the grasshoppers is plants. This may be because plants are more tolerant of genome variation than most other organisms. Even so, only a few cases have been described: euchromatic segments in *Scilla autumnalis* (a member of the lily family, Figure 2.24), heterochromatic knobs in maize (Figure 2.25) and heterochromatic segments in both *Tulipa australis* and the common dock (*Rumex acetosa*).

Rumex acetosa may possess supernumerary segments on chromosomes 1 and 6. The segment on chromosome 6 showed typical mendelian inheritance and a clear geographical localization in areas of the UK where the mean January temperature does not drop below 5°C, i.e. a potential environmental adaptation. In contrast, the distribution of segments on chromosome 1 was ubiquitous and exhibited meiotic drive through the egg, a mechanism which may play a significant role in the maintenance of the polymorphism within the population.

Figure 2.24 Example of *Scilla autumnalis* supernumerary segments. Normal chromosome and homologue with supernumerary segment arrowed. Courtesy of Professor J.S. Parker.

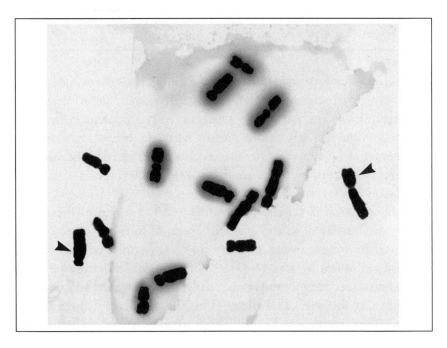

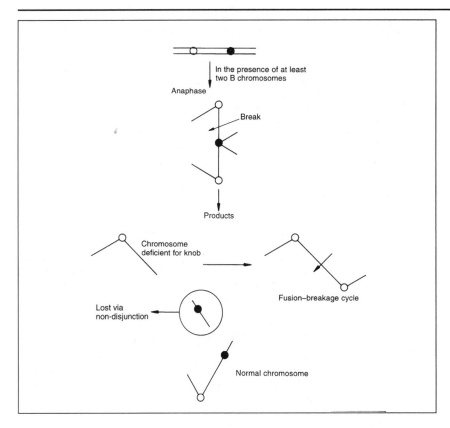

Figure 2.25 Fate of maize knobbed chromosomes in the presence of B chromosomes.

Supernumerary segments and B chromosomes may exist together in the same species. However, only one particular example has been shown of interaction between the two (Box 2.3).

2.10   *Artificial chromosomes*

Construction of an artificial chromosome can be thought of as a first step in testing any hypothesis regarding what is and is not an essential part of a chromosome. The first investigations of this sort started with cloning an origin of replication from yeast. When this was inserted into a cell it was quite capable of independent replication but not stable transmission through cell divisions. Constructed plasmids such as this, which were initially circular DNA, replicated in the cell but did not occur at all in daughter cells. Replication was initiated but not controlled by cell division. The next step was therefore to clone and incorporate the element that controlled chromosome movement during cell division –

Box 2.3   An example of B chromosome and supernumerary segment interaction

Ancient varieties of maize often contain supernumerary segments in the form of heterochromatic knobs that vary in position, size and number between populations. In the presence of at least two B chromosomes, abnormal meiotic events occur at the anaphase of the second pollen grain mitosis.

The A chromosome with the knob fails to separate normally, with the knob appearing to 'stick' the chromosome together. A bridge is formed which eventually breaks between the knob and the centromere of one of the daughter chromatids. This results in one of the pollen nuclei being deficient for the knob and any distal sequences; the other pollen is normal.

The deficient chromosome then has two potential fates, depending on whether it contributes to the zygote or the endosperm, both of which lead to different consequences for the offspring.

When it contributes to the zygote, the broken ends 'heal' and this chromosome is then stable. The genes distal to the knob are deleted, and hence all genes on the maternal segment of this chromosome are expressed, whether recessive or dominant.

Where it contributes via the pollen to the endosperm nucleus, the broken chromosome remains 'unhealed' and enters a breakage–fusion cycle. This is where fusion at the broken ends occurs when the deficient chromosome replicates. This produces a dicentric chromosome and subsequently a bridge at cell division. The bridge breaks at random. These products then go on to produce further dicentrics, which repeat the process in subsequent cell divisions. The ultimate result is a mosaic pattern of expression for dominant genes on the chromosome arm between the centromere and the break.

During this process the B chromosomes undergo non-disjunction. The reason for this peculiar interaction is thought to be a controlling element which, although it recognizes the heterochromatic knobs and B chromosomes as having the same organization, has different effects on their behaviour.

Modern cultivars of maize rarely contain these heterochromatic knobs. This is because they are bred for uniformity, a feature which knobs obviously have the potential to disrupt or alter.

---

the centromere. This was carried out by Clarke and Carbon. Now the plasmids were being inherited accurately in 99% of cell divisions.

The reason that these were circular plasmids was that there was no mechanism present to stop replicative loss of DNA at the unprotected ends if the circular DNA was made linear. Therefore, to turn the circular plasmid into a linear chromosome required the introduction of two telomeres; these were originally cloned from protozoa, but later cloned yeast telomeres were used. Such chromosomes are normally introduced into a yeast cell in the form of a circular plasmid, which is then cleaved at the telomeres into linearity. These first 11-kb yeast artificial chromosomes (YACs) were unstable during cell division, with segregation appearing to be random. Within fairly wide boundaries the stability of YACs during cell division is dependent upon the size. Small YACs are not as stable as larger ones, thus a 200-kb YAC is highly stable during mitosis.

It should be possible using these techniques to create artificial chromosomes for many different organisms, one such being tomato, or for the introduction of genetic material into cells for correcting a genetic lesion, in the form of gene therapy. Once the production of YACs became possible it was realized that they had enormous potential for mapping the genome of many different organisms. Until YACs became available it was only possible to clone approximately 40 kb of DNA into a cosmid vector.

By constructing libraries of YACs (Box 2.4) it is possible to bridge the gap between long-distance mapping and sequencing. The cloning of large numbers of different fragments in YACs can produce a contig library, in which the entire region of the genome or chromosome is represented in overlapping YAC fragments (Chapter 7).

There are, however, specific problems with the use of YACs. Although it is relatively common now to have inserts of 600 kb, it is the very plastic ability of yeast cells to splice foreign DNA into their own genome that is both their remarkable advantage and Achilles' heel. It is quite a common feature of YAC libraries to find 10–60% of cell lines

---

Box 2.4   DNA libraries

A DNA library is an artificial collection of an organism's DNA. The DNA is either cut with restriction enzymes or sheared and ligated into a cloning vector (DNA capable of autonomous replication in either yeast or bacterial cells). The aim is to achieve a single piece of foreign DNA in each vector with enough coverage such that each sequence within the organism is represented at least once.

The three main types of library vector are:

1. YACs (yeast artificial chromosomes). These can hold up to 2 Mb of DNA. Complete YAC contig libraries exist for human chromosomes Y and 21, but libraries are incomplete for the rest of the complement. YAC libraries can be difficult to maintain and are often unstable. Because they can accommodate such large pieces of DNA, they are frequently chimeric, containing two fragments of DNA from different areas of the genome.
2. Cosmids. These are modified plasmids containing some bacteriophage sequences, but also plasmid sequences such as drug resistance (for selection purposes). They can accommodate 35–45 kb of insert. Cosmid libraries exist of most regions of the human genome. They are most useful for building up localized regions of contiguous DNA stretches, for example when searching for candidate gene sequences.
3. Bacteriophage. These are viruses that infect bacteria. Some of their sequences are non-essential in the laboratory environment and can be removed and replaced by foreign DNA (average size 17 kb). The number of bacteriophages required to give a full representation of an organism's DNA is often unmanageable, and they are often used as partial libraries or for subcloning purposes. The size of sequence they can hold is frequently too small to overcome the problems of mapping (i.e. chromosome walking) through regions of repetitive DNA and it is more difficult to build up large contigs in phages than with cosmids or YACs.

Libraries can also be made from RNA. RNA cannot be cloned directly and first has to be reverse transcribed into a DNA copy. These libraries are very useful because they actually represent those sequences within the cell which are being expressed, i.e. proteins produced and are very specific to each cell or organ type.

with DNA spliced in from different parts of the genome under investigation; these chimaeras obviously give a totally incorrect idea of relationships between genes. The introduction of 660 kb YACS into cell lines has been used to show a remarkable stability in inheritance. After 27 cell divisions, half the cells still had the YAC present. In this particular case the artificial chromosome appears as a double minute.

Yeast is an ideal experimental organism for studying artificial chromosomes (Box 2.5).

A novel use for YAC vectors is their introduction into cells of alien species. A 250-kb length of DNA designed to alter expression of a mutant phenotype in mice has been used successfully in such studies by transporting the YAC directly into enucleated blastocysts.

Most recently, interest in artificial chromosomes has been generated by the realization that gene therapy is a practical proposition. Although it would be possible to introduce single genes into cells, there are a number of factors which have to be addressed before gene therapy can be regarded as realistic. The major step is the realization that introducing foreign DNA into cells can in itself lead to problems for the host organism. The use of YACs for gene therapy has been suggested, but ideally the introduced chromosomes would be host in origin.

Experiments with transgenic mice have demonstrated that expression is controllable. This is important because many genetic diseases that would be amenable to gene therapy have tissue-specific affects. For example, cystic fibrosis treatment would need to be primarily targeted at epithelial cells, whereas muscular dystrophy would require the dystrophin gene to be active in muscle cells. The control of expression *in*

Box 2.5 Yeast as an experimental organism

Saccharomyces cerevisiae is a particularly good experimental organism for studying artificial chromosomes for several reasons.

- Although yeast is a simple organism it is eukaryotic, sharing many gene functions with higher organisms.
- The simple metabolism allows flexibility in chromosome numbers that more complicated organisms could not tolerate. Yeast is said to have approximately 18 chromosomes.
- The chromosomes are relatively small, being between 300 kb and 2 Mb, with a total genome size of approximately 15 000 kb. Human Y chromosomes, which are highly polymorphic, vary between 30 Mb and 50 Mb.
- There are relatively few introns; 5000–6000 genes are present with one gene expressed about every 2 kb.
- Errors in cell division are remarkably low for a single-celled organism at 1 in 100 000.
- The functional units of yeast chromosomes are well defined and comparatively short, ranging from 100 bp to 1 kb; human telomeres range from 5 kb to 20 kb.
- As new gene functions are defined they can be put in the context of a well-characterized physiology.

vivo is by promoters and enhancers, which can be introduced with the therapeutic gene. When expression studies of transgenic mice have been made it has been apparent that the presence of prokaryote DNA markedly reduces the level of expression. Perhaps more surprisingly, it has also been found that the lack of introns also reduces expression levels in mice, although not so much in tissue culture systems. It should be remembered here that in this type of system the alien gene and associated carrier DNA is integrated into the host genome in the same way that viruses are.

For stable transmission of large sequences of DNA, which would be the gene of interest and sufficient flanking material to ensure complete expression, YACs are ideal. Production of large amounts of gene containing YACs suitable for use poses a practical problem, but this is not insurmountable. It is important that no loose foreign DNA is simultaneously introduced as active genes from alien species could be devastating. Rearrangements and switching genes on or off could induce tumour growth.

The methods used to introduce the DNA into cells vary from cationic lipids to receptor-mediated entry (Chapter 10). Perhaps the most straightforward method is physically to bombard the cells with particles, although there is a risk of shearing the larger DNA molecules that way. Viruses can be used as vectors, but these are limited in the size of DNA that can be introduced and have the potential to cause problems of their own by integration of the viral material. It is possible to pack up to 150 kb into some of the herpesviruses, but, for example, the cystic fibrosis gene spans 230 kb. Once inside cells, DNA is really quite stable over long periods of time, even without any mechanisms for replication and segregation. Under these circumstances it would be reasonable to assume that repeated treatments would be necessary as the introduced DNA is gradually lost. Although not every cell needs the replacement gene for an adequate expression level of product, there is a lower limit beyond which no alleviation of symptoms will occur. This could be rapidly approached in areas of high cell turnover, such as epithelium or erythroid cells which form blood. It would be better, then, to introduce an artificial chromosome which contains the essential parts for stable inheritance of the gene. These would be an origin of replication, telomeres, centromere and the flanking region of the gene in question.

3 Chromosome identification

Summary
Having described the basics of how DNA is packaged into a chromosome and the different elements contained within the chromosome, the next step is to define the chromosomal constitution of a cell and the identification of individual chromosomes, usually in the form of banding studies.

There are several reasons why it is necessary to analyse chromosome structure in depth, apart from the requirement to define the exact chromosomal constitution of an organism.

- Mutations such as deletions of chromosome segments or additional chromosomes are often associated with human genetic disease. The analysis of these gross chromosomal changes is often the first step in understanding the genetic nature of the disease and narrowing down the position of the genes responsible.

  The identification of marker chromosomes in tumours can often give an indication of how the disease is progressing and ultimately determines the treatment strategy.
- Banding studies aid the mapping of genes to chromosomes.
- The chromosomal constitution of somatic cell lines can be defined.
- In plants, the chromosomal constitution of interspecies hybrids, alien addition lines and somatic fusion products can be identified.

- Evolutionary relationships between species can be inferred at a gross level by comparison of banding patterns.

There are many different methods of identifying chromosomes; which method is most appropriate depends on the species and the information required. The most commonly used methods of chromosome identification (banding) will be discussed here.

3.1 The karyotype

The readily identifiable chromosome structures have already been outlined in Chapter 2. To analyse the chromosomes in a cell, the chromosomes are counted and arranged into pairs. Differences in arm ratios (Figure 3.1) and the number of nucleolar organizing regions can be used to differentiate between similar pairs.

The chromosome pairs are lined up, in order of size, with the short arm uppermost and the centromeres aligned, and the pairs are then numbered in order. This is called an idiogram and is shown for barley in Figure 3.2.

When a karyotype is presented diagrammatically it is called an

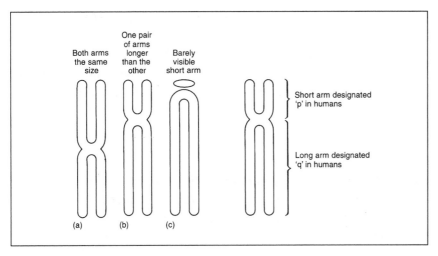

Both arms the same size

One pair of arms longer than the other

Barely visible short arm

Short arm designated 'p' in humans

Long arm designated 'q' in humans

(a)   (b)   (c)

Figure 3.1 Terms applied to differences in size of chromosome arms: (a) metacentric, (b) acrocentric, (c) telocentric.

Figure 3.2 Metaphase spread of barley chromosomes, $2n = 2x = 14$ plain stained with Giesma with karyotype displayed below.

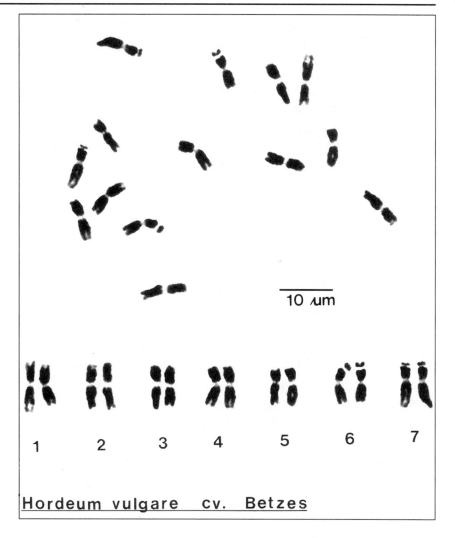

Hordeum vulgare cv. Betzes

idiogram. Each organism, if it has been studied cytogenetically, has a defined standard karyotype, which is a point of reference for mutation studies. But beware, it is not always the smallest chromosomes that are listed last, as can be seen in the barley karyotype. And in humans, chromosome 21 is actually the smallest, not 22. Also, not all organisms are diploid.

3.1.1 Notation

n is the haploid gametic number, i.e. the number of chromosomes in egg or sperm. $2n$ is the number of chromosomes in zygotic cells, in which there are two sets of homologous chromosomes: one maternal, the other paternal.

x defines a basic chromosome set and indicates the ploidy level of the organism. $2x$ indicates that the organism has two basic sets of chromosomes. $4x$ indicates that the organism has four basic sets and is a tetraploid.

In other words, $2n$ is not necessarily the same as $2x$.

Humans are diploid ($2n=2x=46$, where $x=23$), as are Gorillas ($2n=2x=48$, where $x=24$). (These two examples do not define the sex chromosomes.)

However, many organisms, particularly plants, are polyploid, having more than two basic sets of chromosomes, for example bread wheat is hexaploid ($2n=6x=42$, where the $x=7$), whereas potato is tetraploid, ($2n=4x=48$, where $x=12$).

Chromosome sizes and numbers vary dramatically between different organisms and even between members of the same genus. Banding patterns can be produced on most mammalian chromosomes, and this is a way of further refining the description of the cell. The classic example of this is humans, banding patterns being regularly studied as a means of diagnosing some diseases. These, again, are standardized and described using a shorthand set of symbols called the Paris nomenclature:

1. total number of chromosomes;
2. sex chromosome constitution described;
3. any abnormalities described in terms of whole chromosomes or abnormal band patterns.

For example, a normal male is $2n=46$, XY; a normal female is $2n=46$, XX; a female with Down's syndrome is $2n=47$, XX+21.

Similar nomenclature systems exist for most studied organisms such as rodents and cattle, but not for plants.

The more common methods of banding chromosomes will now be described.

3.2 Chromosome banding

The earliest techniques stained chromosomes uniformly; unfortunately, this only allowed a few chromosomes of unusual size or shape to be identified unequivocally. And even if a chromosome can be identified, if it is metacentric then it is not possible to distinguish between the p and q arms.

Unfortunately for geneticists the chromosomes of very few species, unlike the dipteran polytene chromosomes, have a banding pattern built in. For most chromosomes in most species the banded karyotype has to be induced artificially.

Banding of chromosomes normally takes place at metaphase (2n, 4C), but some applications can be used to band chromosomes during prophase when the chromosomes are much longer and can reveal more information.

3.2.1 G-banding (Box 3.1)

Because G-banded preparations contain many bands, this is usually the method of choice for the investigation of chromosomes, especially in clinical cytogenetics. Not all species have chromosomes which can be G-banded; generally, its usefulness is confined to the vertebrates, although G-banding has been demonstrated in some teleosts and amphibians and all reptiles, birds and mammals. Where G-banding is possible, the high information content has allowed accurate phylogenetic analysis between species. Very little success has been achieved with G-banding of plant chromosomes.

Methanol–acetic acid fixation is generally found to produce chromosomes that are morphologically best suited to G-band analysis. This process removes the vast majority of the histone proteins. Therefore, it can be inferred that histones are not important in the banding process. Similarly, since it is necessary to digest part of the non-histone protein of the chromosomes to induce G-bands, it is important that the non-histone protein is not removed by the fixation process. Acid hydrolysis of the DNA, which depurinates it, also inhibits G-banding. Thus, to stain adequately, both protein and DNA components of the chromosomes are required. Although Giemsa is a complex stain based on methylene blue and eosin, it is a precipitation reaction that stains the chromosomes.

In diagnostic cytogenetics, most analysis is carried out on cells with approximately 400 Giemsa bands (G-bands). For more precise work, particularly in cancer cytogenetics, tissue culture techniques are used to prevent the chromosomes condensing as tightly and a banding pattern of approximately 2000 G-bands can be produced on these prometaphase chromosomes. Each chromosome is divided into regions according to the most prominent bands. Each of these can be further subdivided according to the total number of visible bands. These are

Box 3.1 G-banding technique (Figure 3.3)

Figure 3.3 Low-resolution G-banding of a human female metaphase. The X chromosomes are marked.

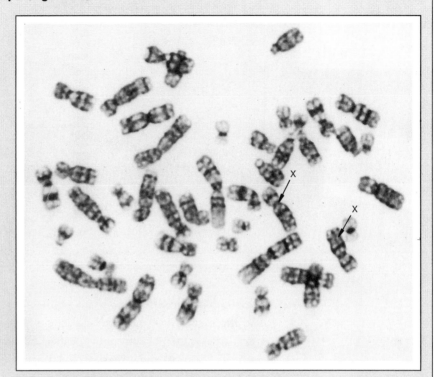

G-banding of chromosomes is a simple two-stage process involving a pretreatment followed by staining with Giemsa, Leishman's stain or Wright stain.

The commonest chromosome pretreatment is enzymatic digestion (Figure 3.4) using trypsin, although it is possible to induce bands using any treatment designed to damage the protein content but not the DNA of the chromosomes. Such treatments include pronase, which is a mixture of proteolytic enzymes, or thermolysine, which digests the proteins and protamines on the amino side of leucine, isoleucine and valine. In contrast, trypsin digests the amino acid bond between lysine and arginine and any other amino acid. High pH can also induce bands by dissolving protein in a systematic manner.

Figure 3.4 Progressive nature of enzymic digestion of chromosomes. This illustration demonstrates the effect of the digestion on the integrity of chromosome structure. At the top of the picture digestion is less than at the bottom. As enzyme action continues, removal of structural protein results in gradual reduction of chromosome rigidity. This causes the chromosomes to collapse and take up a greater area.

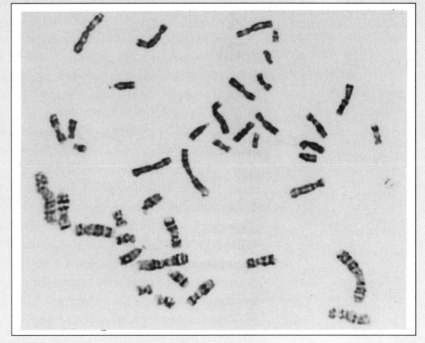

Figure 3.5 Diagram of the human X chromosome, showing band notation and fragile X site.

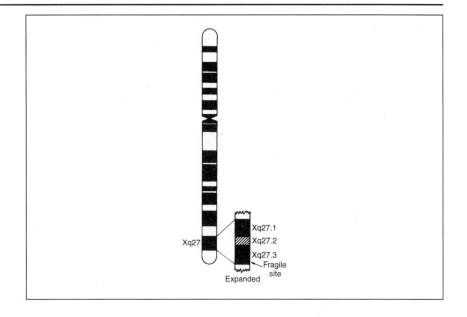

used to describe accurately breakpoints in chromosome rearrangements and the localization of gene sequences. For example, a well-characterized form of mental retardation called Fragile X syndrome is associated with a chromosome break at the point Xq27.3 (long arm, band 2, primary sub-band 7, subdivision 3) (Figure 3.5).

Why are chromosomes divided into light and dark bands and what is their significance? Chemically the dark-staining regions of the chromosomes are 3.2% richer in A and T, but it is unlikely that such a small chemical difference would account for the banding pattern, and it seems reasonable to assume that it is a structural or conformational difference that defines the bands.

It has been proposed that the G-bands of chromosomes are a manifestation of differences in the replication of the chromosomes, with the light bands replicating first, followed by the dark bands. In some rodent cells a short break occurs during replication, leaving an open replication fork. This has been used as an argument in favour of the theory that most breaks and translocations occur at the interface between light and dark bands. This clean break in mid S-phase of replication has yet to be shown in human cells. The physical changes that occur in G-banding can be seen in electron micrographs, in which the surface detail corresponds to the G-bands.

A technique which may be confused with conventional G-banding is the G-11 technique (Box 3.2).

Box 3.2 G-11

This technique is so called because the protein content of a chromosome is dissolved by an enzyme-free solution at pH 11. The technique can be used to differentiate between the chromosomes of different species in rodent/human/ hybrid cell lines. In such preparations the human chromosomes will stain pale blue whereas the rodent chromosomes stain magenta. The exact mechanism underlying this process is still incompletely understood.

3.3.2 C-banding (Box 3.3)

C-banding of chromosomes is used to define regions of constitutive heterochromatin. This type of DNA often contains highly repetitious sequences, such as satellite DNA, and is thought to be genetically inert. The distribution of C-bands can vary considerably from species to species. Unlike G-banding, where phylogenies are reflected in the level of G-band conservation, C-banding is very often species specific or even individual specific. Because of the difficulty in G-banding some plant material, the use of C-banding has become quite important for chromosome identification in plants.

In human chromosomes, C-banding will reveal the centromeres of all the chromosomes, but also the heterochromatin of chromosomes 1, 3, 9, 16 and Y. Only the Y in humans has heterochromatin that is not centromeric, being found at the telomeric end of the long arm.

The normal resolution produced by C-banding metaphase chromosomes can be greatly enhanced by using early metaphase chromosomes which have not completely contracted, or by the use of sophisticated microscopy, such as epi-illumination. Epi-illumination requires that the

---

Box 3.3 C-Banding

C-banding is almost invariably induced by treatment with hydrochloric acid followed by either sodium hydroxide or, more commonly, barium hydroxide treatment and incubation in a warm salt wash and finally Giemsa staining.

The exception to this is when C-banding is sequentially applied after G-banding. This method utilizes an observation from cytogenetics that if G-banding fails then reprocessing the slides for G-banding will occasionally result in C-banding.

The nature of action of the chemicals in C-banding is well documented. The hydrochloric acid treatment depurinates the chromosomes without breaking the phosphate–sugar backbone of the DNA. This is halted before all of the DNA is degraded. The subsequent alkali treatment denatures the DNA, which aids solubilization; incubation in the warm salt wash breaks the sugar–phosphate backbone and DNA fragments pass into solution. Staining has been shown to be reliant on retention of both DNA and proteins (which have yet to be isolated and characterized). This results in the highly compacted heterochromatin being stained with Giemsa as it is the only part of the chromosome that retains both components. The structural basis of C-banding has been well demonstrated by use of the electron microscope, under which the C-band is shown to be more compacted than the lightly stained areas.

Figure 3.6 C-banded human metaphase chromosomes. The very large chromosome 9 (marked) is a normal polymorphic variant. C-banding is a method that can be used to show whether such variants are due to heterochromatin or of clinically significant euchromatin. Photograph courtesy of S.C. Rooney.

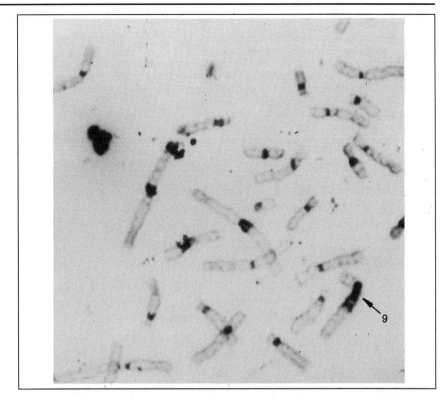

illumination of the object falls on the same side of the object as that from which it is viewed. This has revealed a distinct banding structure within the C-band of human chromosome 9. Bands revealed in such a manner have been termed Ce-bands (Figure 3.6).

3.2.3 Fluorescent banding (Box 3.4)

The very first method of chromosome staining that allowed individual chromosomes to be identified was fluorescent Q-banding. The banding pattern produced is generally similar to that of G-bands. Now more sophisticated cocktails of fluorochromes are available which bind to chromosomal regions with specific base contents.

The mechanism of Q-banding is quite straightforward compared with other banding techniques. The intercalating agent preferentially fits between A and T residues while simultaneously enhancing the fluorescence in these positions. Evidence suggests that the associated proteins may also have an affect, as mouse centromeric constitutive heterochromatin contains a large proportion of AT-rich satellite DNA that fluoresces poorly with Q-banding. Many of the fluorescent dyes can be

Box 3.4 Fluorescent banding

The traditional method of Q-banding utilizes quinacrine dihydrochloride; this molecule intercalates between the DNA and fluoresces when illuminated with UV light.

Fluorochromes which can be used as DNA intercalating stains to produce fluorescent bands come in many shapes and sizes. Some of the alternative stains are acridine orange, ethidium bromide, propidium iodide and Hoechst 33258. As a generalization it can be said that most fluorescent DNA stains are based on a chemical structure of multiple benzene rings.

Multiple fluorochromes used simultaneously on the same metaphase spread have made it possible to enhance the banding pattern by increasing the apparent contrast. For example, actinomycin D can be combined with DAPI (4, 6-diamidino-2-phenylindole). This technique preferentially stains AT-rich regions, often associated with heterochromatin. It is thought that the areas rich in A and T preferentially bind DAPI and the fluorescences alter with the counterstain actinomycin D. Other counterstains that can be used are chromomycin A3 and distamycin A (DA) which visualize GC-rich areas. Many other combinations exist, the use of which can be tailored to specific requirements. For example, when it is necessary to identify the origin of double satellite material in human metaphase spreads, DA/DAPI staining can be quite informative as the satellite material of chromosome 15 is preferentially fluoresced rather than any of the other acrocentric chromosomes (Figure 3.7).

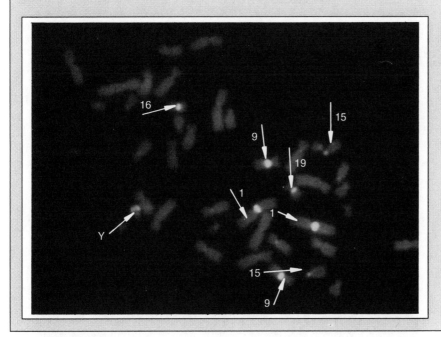

Figure 3.7 DA/DAPI staining of human metaphase chromosomes. The marked chromosomes are the ones most easily distinguished using this technique in humans.

shown to intercalate into DNA in solution and some, such as ethidium bromide, can be used to detect DNA on polyacrylamide or agarose gels, on which the concentration may be very low indeed. The relatively low toxicity of quinacrine dihydrochloride allows it to be used in cell culture without killing the cells. Under these circumstances intercalation into chromosomal DNA undergoing replication can inhibit and alter the action of the spindle in nuclear division as well as inhibiting contraction of heterochromatin.

The major single advantage of quinacrine dihydrochloride and other fluorochromes is that the bonding between dye and DNA is electrostatic rather than covalent, and as such is easily reversible. Consequently, it is possible to Q-band a metaphase spread to identify all the chromosomes, wash the fluorochrome out and then restain with one of the more specialized stains that does not allow for accurate identification.

The drawbacks of fluorochrome banding are, however, quite important for routine diagnostic purposes. The background fluorescence can be considerable if the preparations are not very clean. As the image is

---

Box 3.5   R-Banding

R-bands can be produced either by incubating chromosomes in a hot saline buffer, followed by staining in Giemsa, or by staining heat-treated chromosomes directly with acridine orange (Figure 3.8). In the latter case the acridine orange will fluoresce most intensely in the opposite bands to those which fluoresce with quinacrine dihydrochloride – hence reverse bands.

This technique is particularly use-ful in diagnostic cytogenetics when examining telomeric deletions which with G-banding are normally light staining and difficult to quantify.

The nature of the production of R-bands is thought to be related to the base content of DNA. G-bands are AT rich, and AT-rich DNA denatures at a lower temperature than GC-rich R-bands. After the heat treatment the G-bands would be denatured, but the R-bands are still double stranded, and the nature of the staining reflects this situation. This is borne out by the fact that acridine orange differentially stains single-stranded and double-stranded DNA, a result backed up by the observation that R-bands can be produced by heat treatment followed by application and detection of a monoclonal antibody specific for double-stranded poly-dG and -dC nucleotides.

Figure 3.8   Human metaphase chromosomes stained with acridine orange and fluoresced with UV light. Acridine orange produces a pattern that is sometimes called reverse banding. It must be appreciated, however, that dark G-bands are still dark and light G-bands fluoresce brightly. Courtesy of S.C. Rooney.

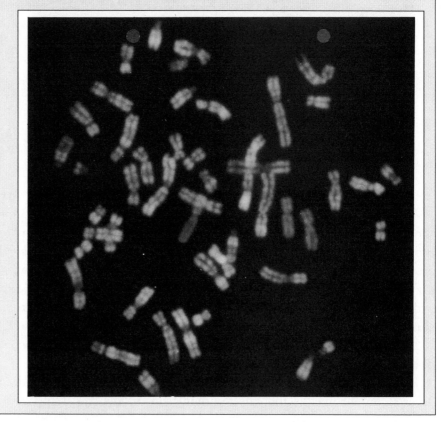

Box 3.6 CpG islands

These are also referred to as HTF islands (*Hpa*II tiny fragment) and methylation-free islands (MFIs).

In general, vertebrate DNA is highly methylated (cytosine bases changed to 5-methylcytosine via the action of DNA methyltransferase). However, approximately 1% of the genome is composed of characteristically non-methylated sequences which show strong correlation with the presence of genes. This is particularly true of mammalian genomes.

These sequences can be cloned by cutting genomic DNA with the methylation-sensitive restriction enzyme *Hpa*II. The small fragments produced are methylated and relatively rich in CpG doublets. There is no consensus sequence, but G/C boxes (GGGGCGGGGC and closely related sequences) are usually present.

In the genome, the CpG sequences occur as discrete 'islands', usually 1–2 kb in length. They are associated with the 5' ends of 50–60% of mammalian genes, including all housekeeping genes and a large proportion of tissue-specific sequences. The CpG sequences are not restricted to 5' flanking regions, but tend to extend downstream into the transcribed sequences, usually the first and second exons.

Over 80% of CpG islands are in the 45% of the genome which comprises R-bands. These are genetically distinct areas of the chromosome, as differentiated by banding studies, which are less condensed and early replicating. CpG islands themselves have been shown to have a distinct chromatin structure. Histone H1 is present in very low amounts, histones H3 and H4 are highly acetylated and nucleosome-free regions are present, resulting in a less condensed packing arrangement.

It has been suggested that this packing arrangement may provide increased access to transcription factors. Also, certain transcription factors, such as Sp1, have a consensus binding site that contains a CpG island and is G + C rich. Hence, CpG islands may represent preferred sites of interaction between DNA and these DNA-binding proteins.

The exact nature of the interaction between CpG islands and transcription has yet to be resolved, as has the relationship between the presence of CpGs and genes. However, CpG islands certainly represent important landmarks within the vertebrate genome.

formed from a radiant fluorescent object, both glare and diffraction from within the object and the mounting medium make it very difficult to produce a sharp image. As a result of this, high-resolution Q-banding is not practicable.

Other drawbacks of this technique are the very high cost of equipment required for fluorescent microscopy; the necessity, because the fluorochromes are bleached by the UV light, to take a photograph of the fluorescent image and to perform the analysis on this; and the toxic nature of the chemicals used, which must be handled with great care.

3.2.4 R-banding (Box 3.5)

R-banding is also referred to as reverse banding.

One intriguing finding from R-bands is that they contain all of the housekeeping genes of the cell and about half of the tissue-specific genes, are rich in GC and have CpG islands (Box 3.6). G-dark bands (the reverse), on the other hand, are rich in AT and have no CpG islands, implying a structural compartmentalization of the genome.

3.2.5   BrdU staining (Box 3.7)

This can be viewed as a special sort of R-banding as the banding pattern produced is frequently the same. However, the methodology of BrdU staining is very different and more complex. Hence it can be manipulated to produce a much higher level of information.

If BrdU is applied for only very short periods (minutes), it becomes possible to track the process of chromosome replication. For example, it has been shown that in cell lines containing double-minute chromosomes, these chromosomes are uncoupled from the exact timing of replication and replicate with no apparent pattern relative to the rest of the karyotype.

If the pulse is incorporated for one complete round of replication the chromatids can be distinguished from each other after the next cell

---

Box 3.7   BrdU staining

This involves the incorporation of thymidine analogues during part of the S-phase of the cell cycle. Incorporation of bromodeoxyuridine (BrdU), a thymidine analogue, into a chromosome during replication alters the staining properties of chromatin after UV exposure and hot washes in buffered saline solution. Where BrdU has been incorporated Giemsa staining is inhibited, so these areas take up less stain and therefore appear pale.

Although the usual method of BrdU detection is by staining with Giemsa, it is possible to use BrdU antibodies or other similar analogues such as bromodeoxycytidine instead.

The most important factor in BrdU staining is the timing of BrdU incorporation. BrdU incorporation is achieved by either of two related methods, the B-pulse or the T-pulse.

B-pulse

In this method, when cells are growing in culture, just before processing the cells a low-thymidine medium, supplemented with BrdU, is substituted for the original growing medium. At this point any replicating chromosomes will incorporate BrdU. After further processing the areas which have taken up BrdU are pale staining when compared with the non-incorporated areas. In this method BrdU uptake is going on while replication is finishing so that the dark G-bands will contain BrdU and therefore stain lightly, i.e. reverse banding.

T-pulse

The alternative method is the T-pulse method. This is rather more difficult because of the cytotoxic nature of BrdU. The cells are grown in low-thymidine medium supplemented with BrdU, which is replaced towards the end of the cell cycle with BrdU-free medium. Since the early part of the cycle is taking place in the presence of BrdU, the early-replicating parts of the chromosome will be pale when stained. The light G-bands are the first to replicate so the T-pulse method corresponds to G-banding: not reverse banding but replication banding.

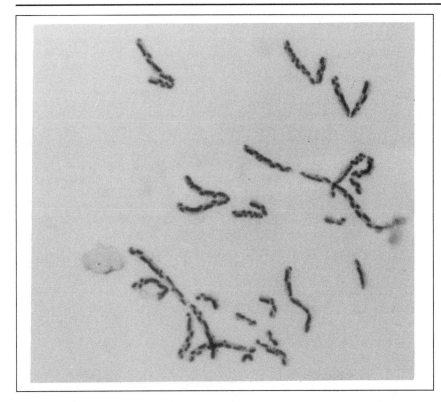

Figure 3.9 Sister chromatid exchange (SCE) demonstrated in a susceptible Chinese hamster cell line EM7. The unusually high rate of SCE is clearly visible. Courtesy of A.N. Harvey, MRC Radiobiology Unit.

division, one being pale and one being dark. If sister chromatid exchange (SCE) takes place then the pattern of staining is referred to as harlequin banding (Figure 3.9).

Because chemical mutagens and radiation cause most damage during DNA replication, SCE measurements can be used to study the result of exposure to mutagens, although it should be noted that ionizing radiation is not as effective at inducing SCE as chemical mutagens because of the transitory nature of radiation.

3.2.6   RE banding (Box 3.8)

This type of banding is produced by restriction enzymes (REs), alternatively called restriction endonucleases.

In general, the REs produce banding patterns similar to C- and G-bands (Figure 3.11). For example, *Alu*I, *Mbo*I, *Rsa*I and *Msp*I all produce C-bands, *Hae*III produces G-bands in primates and *Taq*I can G-band human chromosomes. The same REs are not necessarily equally successful or produce the same type of banding pattern in different species.

Although the primary factor determining RE bands is assumed to be

Box 3.8 Restriction enzymes

There are several types of restriction enzyme (RE), but usually type II enzymes are used for this type of study. REs are strain-specific proteins produced by bacteria which recognize and degrade any foreign DNA entering the cells. These enzymes recognize a specific base sequence (usually 4–7 bp in length) of DNA and introduce a cut into that DNA at a defined place.

The names of REs are codes denoting their origin. The first three letters stand for the name of the bacterium (e.g. *Eco* for *E. coli*) and the fourth letter for the strain (e.g. *EcoR*). If more than one RE is found in the same strain, these are denoted by roman numerals, e.g. *Eco*RI is the first RE from *E. coli* RY13 and *Hae*III is the third RE from *Haemophilus aegyptius*.

RE banding works on the basic premise that, if enough cuts are made in the DNA sufficiently close together, the pieces are small enough to be extracted from the chromatin matrix by a gentle incubation. If sufficient pieces are removed in any given area, the DNA concentration will decrease and consequently stain more lightly than surrounding areas, thus producing a banding pattern (Figure 3.10). RE banding has been used on a wide variety of organisms, mammals, insects, fish, birds and amphibians.

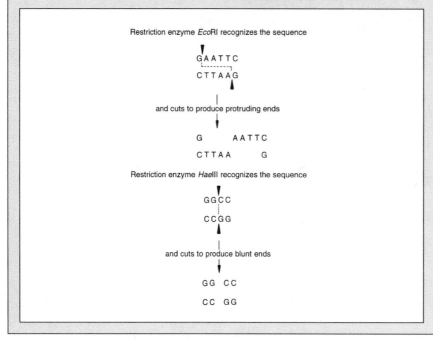

Figure 3.10 Action of restriction enzymes on DNA.

the base sequence of the DNA, other factors also influence whether an RE will cut. These often reflect constraints imposed by the higher order structure of the chromosome.

1. DNA is generally extracted when the fragments are between 200 bp and 4 kb in size. This process is affected to a certain extent by the presence of chromosomal proteins, as a protein digestion stage has been shown in some cases to release further DNA fragments. Also, if before treatment the chromosomes are subjected to DNA–UV–protein cross-linking, inhibition of RE action occurs. There is also domain variability in the rate of RE digestion, e.g. *Hae*III cleavage is more rapid in human R-bands than in G-bands; this is assumed to be because of the different protein

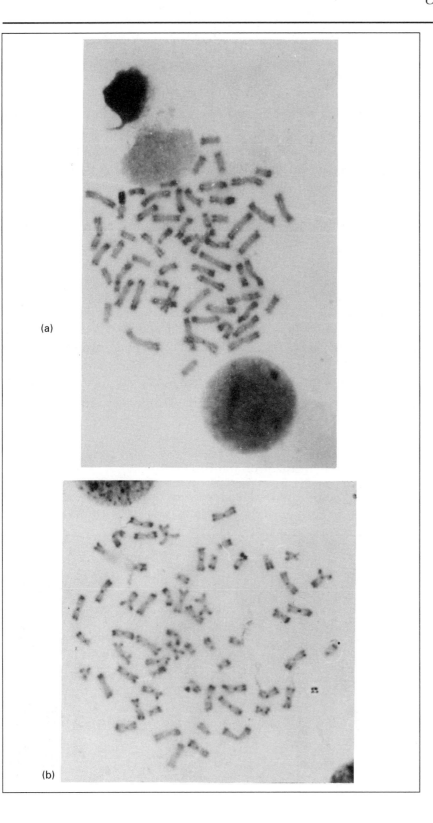

(a)

(b)

Figure 3.11 Metaphase chromosomes of the coho salmon (*Oncorhynchus kisutch*) digested with (a) *Mbo*I and (b) *Ode*I. Reproduced with permission from Lozano *et al.* (1991), *Heredity*.

constitution of R-bands, in which active genes are predominantly located.

2. Some organisms modify cytosine bases to 5-methylcytosine as a protection measure against DNA degradation by the restriction enzymes of other organisms. Many enzymes cannot cut when these modified bases are present.

3. Chromatin conformation also plays a role. For example, *Bst*NI cuts naked mouse satellite DNA into 245-bp fragments, but does not digest mouse chromosomes at all. If 5-azacytidine is used to decondense the chromatin, *Bst*NI will then cut. This enzyme has a higher molecular weight than most of the other enzymes, and it has been suggested that this lack of action is the result of inability of such a large molecule to access the three-dimensional structure of the chromosome.

RE banding may be considered an expensive means of producing C- and G-bands, but it is particularly of use in species which do not band well (amphibia, plants and fish) (Figure 3.12). In many cases RE banding is more reproducible and better defined than conventional C-banding. Other uses of RE banding include evolutionary studies, in which RE digestion can give information about the distribution and conservation of repeated DNA families.

3.2.7   In situ hybridization banding (Box 3.9)

This type of banding is produced by *in situ* hybridization techniques.

Mammalian chromosomes contain a large proportion of clustered

Figure 3.12 Haploid karyotype of *Scilla siberica* after treatment with *Hae*III and proteinase K. C-banding on the left, enzyme digestion followed by Giemsa staining in the middle, with enzyme digestion followed by staining with the fluorochrome propidium iodine on the right. Reproduced with permission from Lozano *et al.* (1991), *Chromosoma*.

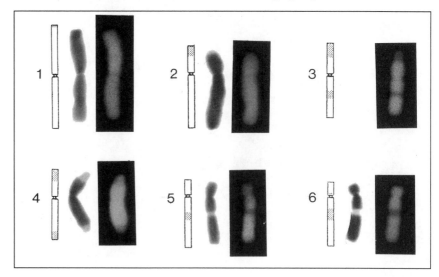

Box 3.9 *In situ* hybridization

This is one of the more common methods for investigating chromosome structure. It allows the detection of a sequence of interest in a chromosome. This can either be in sections of tissue or on the chromosome itself.

Standard chromosome preparations are made on a slide and heated to denature the strands of DNA. If the preparation is rapidly cooled in the presence of formamide, the DNA strands stay separated. The gene or sequence of interest is cloned and has a label incorporated into it, which facilitates its later detection. This sequence is also denatured to make it single-stranded, added to the chromosome preparation and incubated. The DNA in the chromosomes and also the label will naturally reanneal, but the probe and chromosomal sequences compete with each other such that some of the probe will hybridize to the chromosomes. It is possible to detect this later under the microscope after having performed a series of antibody reactions to enhance and detect the label incorporated into the probe.

When the technique was first developed the label incorporated into the sequence of interest was a radioactive isotope, tritium. This was a slow and rather unwieldy technique. The more recent popular alternative to radioactivity is to use immunological methods, whereby the label can be detected using antibodies. These are linked either to an enzyme which can cause a colour reaction when introduced to the appropriate substrate or to a fluorochrome. The former is detected under the light microscope as a coloured precipitate on the chromosome, the latter as a fluorescent signal when viewed under a fluorescence microscope.

This has led to an explosion in the number of labels available and hence the number of different sequences which can be detected at any one time (current count is 12 simultaneously!). Also, the technique is becoming increasingly sophisticated in terms of resolution and complexity of uses.

This technique can be modified so that analysis can take place under the electron microscope (for greater resolution) or on embedded sections to analyse whole tissues, this is often used to examine the expression pattern of a gene. Whole mounts of small organisms or embryos can be used in developmental studies. Increasingly sophisticated confocal microscopes can now also build up three-dimensional images by optical sectioning. Probes can take the form of single gene sequences, whole chromosome segments or total genomic DNA or RNA.

This technique is associated with more than its fair share of abbreviations and acronyms; some examples and explanations are listed below.

- ISH. *In situ* hybridization.
- FISH. Fluorescence *in situ* Hybridization.
- GISH. Genomic *in situ* hybridization. Whole genome DNA is labelled and hybridized to whole chromosome spreads to pick out all the chromosomes of a particular species. Repetitive sequences anneal more rapidly than conserved sequences, and these repetitive sequences often show little evolutionary conservation and hence can be used as a species-specific label, for example in somatic cell lines, fusion hybrids and evolutionary studies.
- CISS. Chromosomal *in situ* suppression. DNA from a single type of chromosome is labelled, prehybridized with repetitive DNA from the same organism to stop cross-hybridization with other chromosomes from the same organism and hybridized to chromosome spreads or tumour sections. This is often used to determine the chromosome constitution in the progression of cancers, in particular solid tumours from which it is difficult to obtain chromosome spreads. It is now also used in prenatal diagnosis, the determination of cell line constitution, definition of chromosome abnormalities and in evolutionary studies, for example to follow the changes taking place in one particular chromosome between related species.
- WCP. Whole chromosome painting probes. As used in CISS above and available commercially for all 22 autosomes and the sex chromosomes of man.
- DIRVISH. Direct visual hybridization. Free linearly extended DNA from nuclei is hybridized with differently labelled single-copy probes to produce a chromosome map, the resolution of which is approximately 1 Mb.

- PRINS. Primed *in situ* labelling. Sequences are labelled on the actual chromosome or tissue using an adaption of the traditional polymerase chain reaction (PCR) technique.
- FICTION. Fluorescence immunophenotyping and interphase cytogenetics as a tool for investigation of neoplasms. Tumour cells often contain a mixture of abnormal and normal cell types. This technique immunodetects the abnormal cells, and subsequent *in situ* hybridization of whole chromosome probes can be directly related to cell type, improving diagnosis of tumour progression.

No doubt there are many more uses waiting to be discovered.

In situ banding exploits the fact that DNA of all organisms contains highly repetitious sequences which are present in the genome in several thousands to hundreds of thousands of copies. These sequences can be dispersed at random throughout the genome or clustered in blocks. If these blocks are large enough it may be possible to detect their presence on chromosomes by *in situ* hybridization. Also, if they are present on all or most of the chromosomes in a distinct pattern, they can be used as karyotyping aids.

repeat sequences, the major component of which is satellite DNA. This is located mainly in the centromeric heterochromatin regions. There is also repetitive DNA interspersed among single-copy DNA. It is divided into two classes: short interspersed repeat sequences (SINEs) and long interspersed repeat sequences (LINEs). The terms 'short' and 'long' refer to the length of the basic sequence motif. Both of these examples have been mapped by *in situ* hybridization. In general, LINEs map to G-bands (G-dark bands) and SINEs to R-bands (G-light bands). These sequences were first isolated and mapped to further the understanding of the eukaryote genome organization. However, a further use has been found: the species specificity of some of these repeat sequences can be used to identify human chromosome fragments in somatic cell hybrids.

In plants, *in situ* banding has a potentially more powerful karyotyping role. Plant chromosomes, in general, are not G-banded to a useful extent. C-banding produces too few bands which are too evenly distributed to be useful karyotyping aids. Many plant chromosomes are morphologically similar and in particular many crop plants, e.g. of the *Brassica*, *Solanum* and *Gossypium* genera, have small genomes and consequently minute chromosomes which are difficult to identify. Using a combination of *in situ* banding, total chromosome length and arm ratios it has been possible to define an *in situ* karyotype of tomato using the TGRI sequence (Figure 3.13).

Since in many plants it is difficult to identify individual chromosomes, aneuploid plants have to be used. These are plants whose genome has been manipulated by the introduction or deletion of one or two chromosomes. Using different aneuploid plants, the number of *in situ* hybridization signals can be compared with the number of

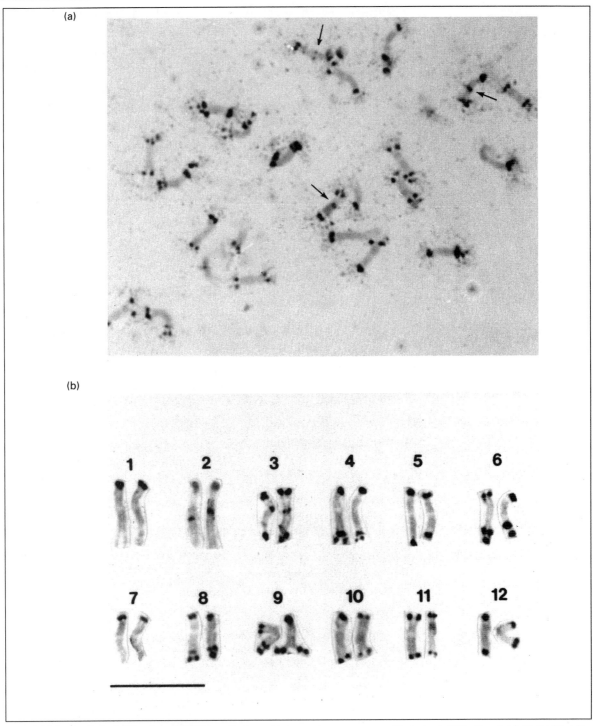

Figure 3.13 *In situ* hybridization of TGRI to mitotic metaphase chromosomes of tomato. (a) Whole cell of plant trisomic for chromosome 3 (arrowed). (b) Somatic karyotype of tomato based on the hybridization of the TGRI probe. Reproduced with permission from Lapitan *et al.* (1989), *Genome*.

a particular chromosome present in each, hence defining which chromosome the sequence came from. The production of these is a complex and time-consuming process, and hence they are not available in many species, limiting the usefulness of this technique.

Similar to their use in the medical field, repetitive sequences in plants can be used to determine the chromosome constitution of somatic hybrids and also evolutionary relationships between different species.

It should be noted that *in situ* banding is far more specialized than the other types of banding described. It requires a great deal of technical back-up, in particular the cooperation of a molecular biology laboratory for all the cloning work and characterization of isolated sequences. Therefore this technique is really only feasible in the larger laboratories and for a very specific purpose, not routine identification.

3.2.8 D-banding

D-banding is produced by the differential sensitivity of chromatin to the enzyme DNase I (or pancreatic deoxyribonuclease I, in full). It is also called DNase I sensitivity banding.

The enzyme DNase I rapidly degrades purified DNA. However, DNA in chromatin is protected to a certain extent by the presence of chromosomal proteins; the rate of digestion, therefore, is dependent, primarily, on the binding of these proteins. It has been shown that active genes are more sensitive to DNase I digestion than those that are inactive. It is therefore possible to use the relative rates of digestion to determine relative differences in chromatin structure. In some organisms, this can be used to produce a banding pattern using the technique of *in situ* nick translation (Box 3.10).

The knowledge that the protein constitution of chromatin changes when genes are activated and the similarity of dark D-bands and light G-bands (Figure 3.14), which mainly contain housekeeping genes, led

---

Box 3.10 *In situ* nick translation

Commercial nick translation DNA-labelling kits are available in which DNase I is combined with DNA polymerase I (DNA pol I). The DNase I introduces nicks at random into the DNA and degrades it. This is followed by the action of DNA pol I, which then reconstructs the degraded part of the DNA. The presence of free nucleotides is required to manufacture the new DNA. If a labelled nucleotide is included, this will also be incorporated into the chromosome at the sites of DNase I action, and it is then possible to detect this labelled section of the chromosome under the light microscope. The types of label and methods of detection are the same as those listed for *in situ* hybridization (tritium, digoxigenin, biotin).

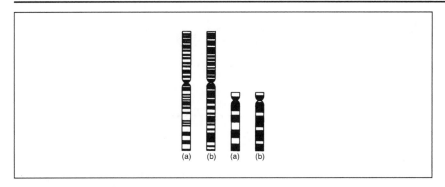

Figure 3.14   A comparison between (a) G- and (b) D-bands of chromosomes from a CHO (Chinese hamster ovary) cell line. Reproduced with permission from Kerem *et al.* (1984), *Cell*, **38**, 496, Figure 4, copyright Cell Press.

to the theory that D-banding identifies functional domains of the chromosome. Interestingly, not all light G-bands correspond to dark D-bands. Hence, D-banding further aids the understanding of the structure and function of mammalian chromosomes. It provides an alternative technique for karyotyping (although only with organisms whose chromosomes G-band) and potentially adds a new dimension to the study of genetic disorders with its identification of functional domains. It is particularly useful in defining the activity of X chromosomes in mammalian females. For example, the inactive X chromosome in *Gerbillus gerbillus* is identical to the active X in G-, Q- and R-banding, but differs in its DNase I sensitivity. Also, it has been shown that not all of this X is inactive and a small area remains active. (See chapter 8 for more detail on sex chromosomes and X inactivation.)

Banding primarily provides a physical map with which to identify individual chromosomes. This, in turn, allows the identification of chromosome mutations within either an organism or a population and also the study of evolutionary links between related species. Recently, the increased sophistication of the techniques and in-depth analysis of the individual banding mechanisms has hinted at a superimposed compartmentalization of chromosome structure. Hence, the use of banding has been extended to help unravel the true nature of this complex organelle.

4 Nuclear division

Summary

The perpetuation of life requires many essential biochemical reactions; however, most importantly, underlying all these is the blueprint or map of all the genes necessary for these processes. The vast majority of cells contain this map in the form of DNA (chromosomes), which must be accurately reproduced in each cell to allow these processes to continue.

Nuclear (or chromosome) division takes two forms:

1. mitosis, in which the chromosomes of the parent cell are accurately reproduced in the two daughter cells;
2. meiosis or reduction division which takes place during gamete formation. Two divisions with only one replication take place so that only the haploid number of chromosomes is passed onto the four daughter cells.

There is a considerable difference between duration of mitosis and meiosis both in the same organism and between different organisms. Examples from plants are given in Table 4.1.

Perhaps the best-known example of the differences in duration of mitosis and meiosis occurs in humans. The average duration of the mitotic cell cycle is 24 h. Meiosis in the spermatocytes takes approximately 24 days. In the female meiotic divisions start before birth, but are arrested at early diplotene (dictyotene) of the first division. Completion is initiated at the start of the menstrual cycle with only one or two eggs stimulated each month. Towards the menopause, eggs stimulated to complete the cell cycle will have been in suspended animation for over 40 years.

Table 4.1 Mitotic cycl
and duration of meiosis
plants

Species	Mitotic cycle time (h)	Duration of meiosis (h)
Triticum aestivum (wheat)	10.5	24.0
Secale cereale (rye)	12.75	51.2
Allium cepa (onion)	17.4	96.0
Tradescantia paludosa	20.0	126.0
Happlopappus gracilis	11.9	24.0
Trillium erectum	29.0	274.0
Lilium longiflorum	24.0	192.0

Data taken from Bennett (1971).

4.1 Mitosis

This is often termed somatic cell division, although it does occur during the primary stages of gamete formation in higher animals (production of spermatogonia and oogonia). It can also occur in the absence of cell division, as in the case of the *Drosophila* embryo, which starts as a coenocyte before differentiating into cells.

The behaviour of chromosomes in the mitotic cycle is easily studied by staining an actively dividing tissue, such as a plant root tip, with a basic dye, e.g. Feulgen, which reacts with the aldehyde groups released from DNA after a mild acid hydrolysis. This dye only stains the DNA.

For most of the cycle the nucleus is in interphase. The chromosomes are present as very long diffuse threads, but are not contracted enough to be visible in the nucleus. It is during this period that RNA transcription and DNA synthesis takes place. A dense RNA body called the nucleolus is present in the nucleus.

In humans, mitosis occupies only 20–60 mins of the 24-h cycle, 6–8 h being taken up by DNA synthesis. In *Vicia faba* (bean), in which mitosis lasts 2 h, G_1 takes 5 h, DNA synthesis 7.5 h and the second gap stage, G_2, lasts 5 h (Figure 4.1).

Mitosis can be divided, for ease of explanation, into four stages. Each of these will be described for *Hypochoeris radicata* ($2n=2x=8$) via photographs and an idealized cell diagram.

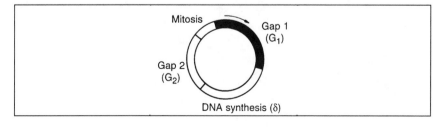

Figure 4.1 The cell cycle.

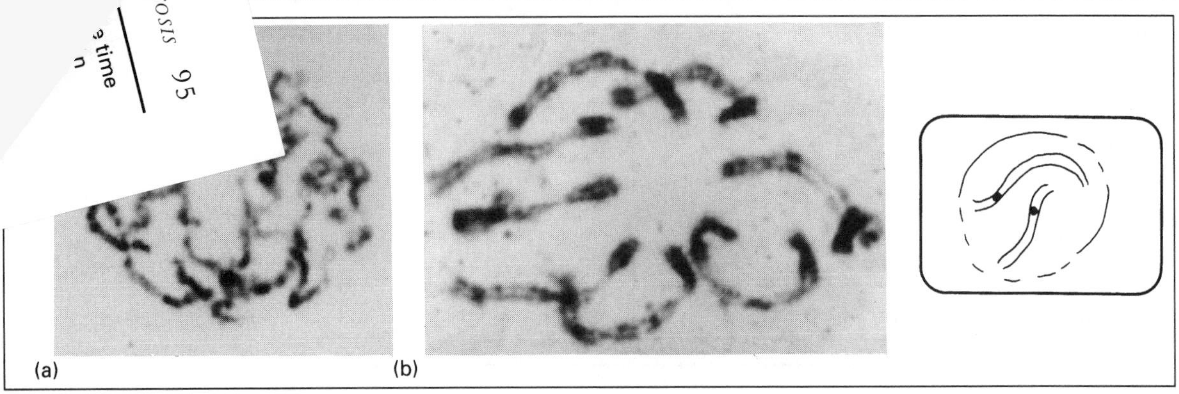

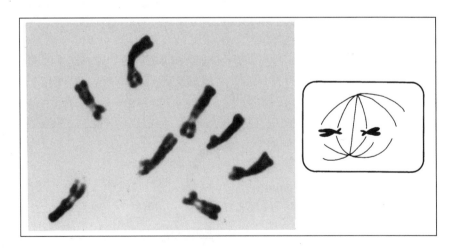

Figure 4.2 Prophase in *Hypochoeris radicata*. (a) Early prophase. The chromosomes appear as diffuse threads with indistinct morphology, although it is occasionally possible to make out their double-stranded nature. (b) Late prophase. The chromosomes are now well separated and distinctly double-stranded and features such as centromeres are visible. Photograph courtesy of Professor J.S. Parker.

4.1.1 Prophase (Figure 4.2)

The chromosomes start to coil up and become visible at the light microscope level. Each chromosome appears to consist of two threads or chromatids held together at the centromere. The nucleolus breaks down and the mitotic spindle apparatus starts to form. The nuclear membrane starts to break down and large organelles are cleared from the central region of the cell/spindle.

4.1.2 Metaphase (Figure 4.3)

The chromosomes are now free in the cytoplasm. The spindle fibres form a radial array, some of which attach to the centromeres of the chromosomes. The chromosomes are maximally contracted and align on a flat plane midway between the poles on the spindle equator (metaphase plate).

Figure 4.3 Metaphase in *Hypochoeris radicata*. Photograph courtesy of Professor J.S. Parker.

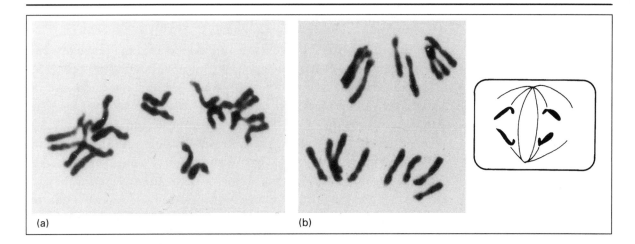

The photograph shows all eight chromosomes maximally contracted. The view is of the metaphase plate from above.

4.1.3 Anaphase (Figure 4.4)

The centromeres of the chromosomes are in fact double structures held together by pericentric regions. These regions dissolve and sister chromatids separate and are pulled to opposite poles, so that each pole receives the chromatids of a complete set of chromosomes.

Figure 4.4 Anaphase in *Hypochoeris radicata*. (a) Early anaphase with the attractions between the chromatids beginning to lapse and poleward movement starting. (b) Late anaphase, when chromatids are almost lost at the poles. Photograph courtesy of Professor J.S. Parker.

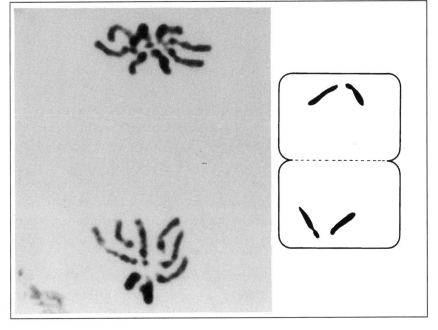

Figure 4.5 Telophase in *Hypochoeris radicata*. Photograph courtesy of Professor J.S. Parker.

4.1.4   Telophase (Figure 4.5)

The spindle breaks down, the nuclear membrane envelops the chromosomes and the nucleolus reforms within the nucleus. The chromosomes uncoil, become diffuse and are no longer visible at the light microscope level. Cytokinesis (cytoplasm division) takes place with the formation of a cell plate in plant cells and an indented cell furrow in animal cells. The cellular organelles such as the mitochondria are partitioned between the two new daughter cells. The cells then enter a 'resting phase', during which more organelles are synthesized to reconstitute the normal number for the cell. DNA is also replicated from the chromatids to form full chromosomes and restore the DNA amount to that of the diploid level.

The photograph shows the chromatids at the poles, becoming more diffuse, entering into interphase.

4.2   The spindle apparatus

Chromosome division is dependent upon the formation and action of the spindle. This is a bipolar structure radiating from a central position in each pole to form an extensive network throughout the cell. It is primarily composed of microtubule protein units. These protein units form fibres with a distinct polarity. The pole end is designated 'minus', whereas the opposite end is designated 'plus'. These microtubules do not act alone, and it is becoming increasingly apparent that the dynamics of the spindle is driven by many associated microtubule protein motors.

The following observations are based on experiments conducted on mitosis, however similar structures and processes occur in meiosis. Variants obviously exist between the two types of nuclear division and also between species.

Where do the spindle fibres (Figure 4.6) attach to the chromosome and what is the nature of this attraction? The prime candidate is the kinetochore, a protein structure found within the centromere region of the chromosome (the structure of which was described in Chapter 2).

Addition of anticentromere antibodies (against the CENP proteins) at any point 3 h before the onset of mitosis inhibits some part of the process of assembling the trilaminar structure of the kinetochore. The

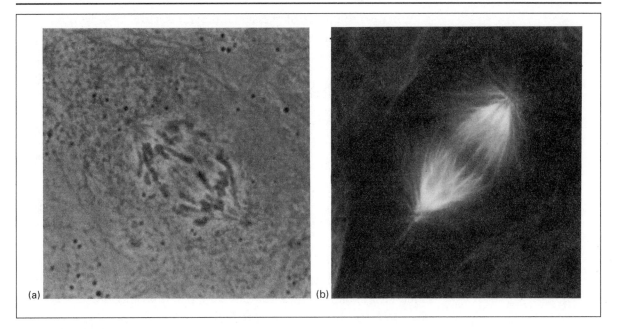

(a)

(b)

Figure 4.6 (a) Phase-contrast image of a mid-anaphase cell from a porcine kidney cell line. (b) Immunofluorescence with antitubulin antibodies to show distribution of the mitotic spindle fibres. Reproduced with permission from Gorbsky (1992), *BioEssays*, **14**.

spindle fibres can still bind, but the chromosome is unable to move along them, indicating that the processes of microtubule binding and chromosome movement are distinct events. Addition of the anti-centromere antibodies 2 h before the onset of mitosis has a different effect: the trilaminar structure of the kinetochore assembles, but binding of the microtubules is unstable. It is thought that this instability is due to problems in condensation of chromatin around the kinetochore structure. The chromosomes can align on the metaphase plate, but all further progress is blocked. This indicates that the kinetochore, in addition to having an important role in microtubule binding, is also involved in some part of the cell signalling pathway that directs chromosome movement. It is thought that, although the addition of anti-centromere proteins affects kinetochore function, the main action is through interfering with the binding and action of a whole range of associated microtubule motor proteins.

Although the structure to which the spindle fibres bind is known, little is known about the nature of this binding, in particular the initial 'capture' of the chromosome. It is thought that the first hits are unstable, particularly if only one of the kinetochores of a pair of sister chromatids is attached. The chromosomes oscillate vigorously back and forth until they eventually attain a stable position on the metaphase plate, equidistant from both poles. This process is often called congression.

What is the nature of the signal that causes the attraction between kinetochores of sister chromatids to lapse? Recent evidence links this to a transient rise in calcium ions (Ca^{2+}). This inactivates maturation-promoting factor (MPF), which is a protein kinase complex containing various proteins, including cyclin B. MPF induces and maintains cells in mitosis via degradation of cyclin B. A reduction in MPF is thought to initiate a cascade of enzyme events that eventually leads to sister chromatid disjunction.

Once attraction has lapsed, how do chromatids move to the poles? Initially it was believed that chromosomes were attached to fibres, which were then reeled into the poles. This is 'the traction fibre' hypothesis developed by Klein and Van Beneden as long ago as the late nineteenth century. However, recent studies using fluorescent probes attached to tubulin have shown that most of the movement of the chromatid pairs during mitosis is the result of movement of kinetochores along microtubules and that this movement can take place in both directions (Figure 4.7).

Microtubules grow and contract via polymerization and disassembly of GTP-bound tubulin units, the overall process of which has been shown to be governed by motor proteins. These are members of the

Figure 4.7 Potential forces involved in moving chromosomes during prometaphase. (A) Motors moving towards the pole end of the microtubules. (B) Disassembly of the microtubules. (C) Plus end-directed motors may push the chromosomes away from the poles. (D) Chromosome arms may be subject to an ejection force which drives the chromosomes away from the poles. (E) Traction force caused by the slow movement of microtubules poleward. At anaphase, forces (A) and (B) predominate. Reproduced with permission from Gorbsky (1992), *BioEssays*, **14**.

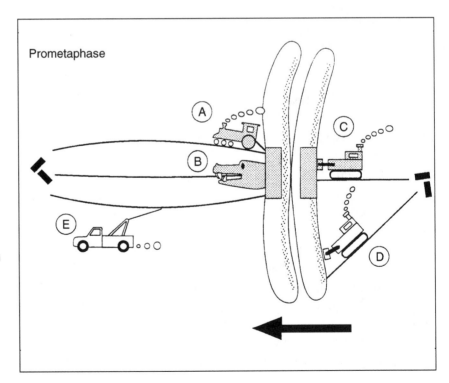

Prometaphase

kinesin and dynein families. Their role is only just beginning to be analysed, and this may take a long time as the kinesin family in *Drosophila* has been shown to have between 30 and 35 members.

Several loss-of-function mutants which have been discovered in *Drosophila* are beginning to be used to answer some of the questions surrounding the motor proteins.

The claret or *ncd* (non-claret disjunctional) motor protein acts early in meiosis and has roles in spindle assembly and bipolarity. It also plays a role in chromosome capture by the spindle and polar body formation in early mitosis.

nod (no distributive disjunction) mutants act only on chromosomes that have not undergone recombination, indicating that, like chiasmata, it acts to oppose the forces moving the chromosome to the poles and maintain the position on the metaphase plate.

Both proteins share domain homology with kinesin, a motor protein that moves the chromatid towards the plus end of the microtubules. More kinesin proteins have been found in other organisms, indicating their widespread distribution. Further research will define their exact roles in both mitotic and meiotic chromosome movement.

4.3   Meiosis

This is reduction division and occurs during gamete formation to ensure that each gamete receives only one set of chromosomes. Hence, when two (haploid) gametes fuse to form the zygote (diploid) there are again two sets of chromosomes.

In essence, meiosis consists of two cell divisions, but only one division of the chromosomes. It is also in meiosis that exchange of genetic material between maternally and paternally derived chromosomes (usually called recombination, although in the strictest sense recombination only occurs when the paternal and maternal chromosomes differ in content, which would not be the case with a pure inbred line) and independent assortment takes place. The different stages of meiosis will be described, followed by further details on the mechanisms and structures behind chromosome pairing and recombination.

The stages of meiosis in *Lilium* cv. Enchantment will be illustrated by photographs and by idealized cell diagrams.

Figure 4.8 Mitotic metaphase spread of *Lilium* cv. Enchantment ($2n = 2x = 24$). Photograph courtesy of Professor J.S. Parker.

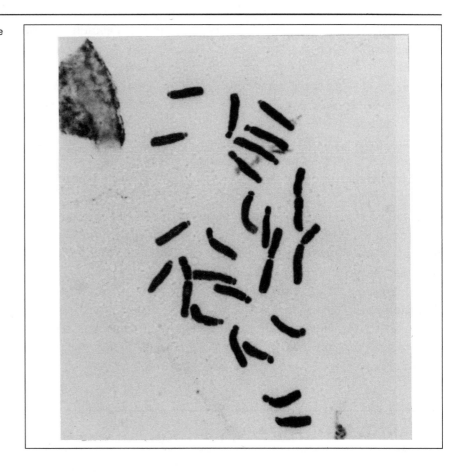

The standard karyotype ($2n{=}2x{=}24$), in the form of a mitotic metaphase spread, of *Lilium* cv. Enchantment is shown in Figure 4.8.

4.3.1   First meiotic division

Prophase

In preparation for meiosis, the chromosomes replicate but the products of replication cannot be visualized during prophase. Prophase is a fairly lengthy process, which can be subdivided into five sections.

Leptotene (Figure 4.9)

The chromosomes, when they appear, look like threads, or often a beaded string, owing to the presence of chromomeres. Each chromosome appears to be single without evidence of replication into

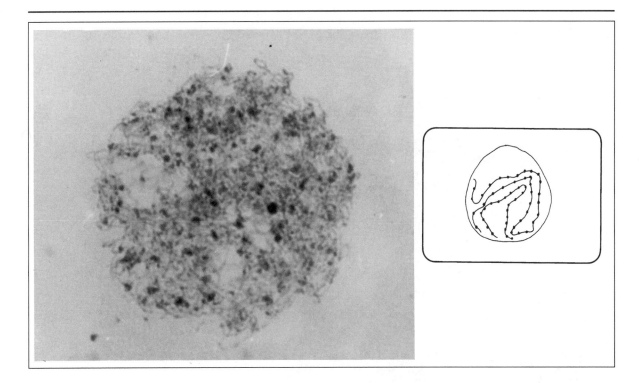

chromatids. It is difficult to define and follow the progress of individual chromosomes within the nucleus.

Zygotene (Figure 4.10)

Homologous chromosomes are attracted and pair (see below). This
process commences at well-defined sites in many species. In animals
this is often near the telomeres, but in plants can start at many sites. A
structure called the synaptonemal complex (SC) (described later) starts
to form between paired homologous segments and chromosomes, but
this is not visible at the light microscope level.

It is still difficult to differentiate individual chromosomes from the
seemingly tangled 'ball of string'.

In the mammalian male, the X and Y chromosomes, after pairing,
undergo a different pattern of coiling and are temporarily spatially
isolated in the cell in the form of a sex vesicle during prophase. One of
the reasons put forward for this is that lack of pairing can prevent
successful meiosis in animals. Since the X and Y do not pair over their
whole length, this has the potential to destabilize the meiotic process
and therefore they are temporarily isolated in the cell during the crucial
period.

Figure 4.9 Leptotene in
Lilium cv. Enchantment.
Photograph courtesy of
Professor J.S. Parker.

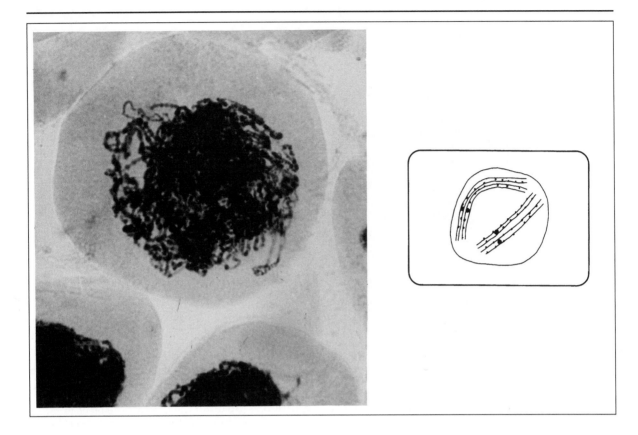

Figure 4.10 Zygotene in *Lilium* cv. Enchantment. Photograph courtesy of Professor J.S. Parker.

Pachytene (Figure 4.11)

Pairing is now completed and the chromosomes contract further. Homologous chromosomes are closely associated (now called a bivalent). Genetic crossing over occurs with the physical exchange of DNA between maternal and paternal chromosomes. It is at this stage that the nucleolus often disperses.

As long ago as 1937, Mather demonstrated that chiasma frequency per bivalent is directly related to chromosome length. Long chromosomes may have several chiasmata, but to ensure proper segregation at anaphase I all bivalents must have at least one chiasma.

Diplotene (Figure 4.12)

Chromosome contraction continues and attraction of homologues lapses. Each chromosome is now clearly visible and acts as if it is repulsing its closely paired homologue, but they are held together at the site of crossing over by visible structures called chiasmata. Sister chromatids are still attracted to each other.

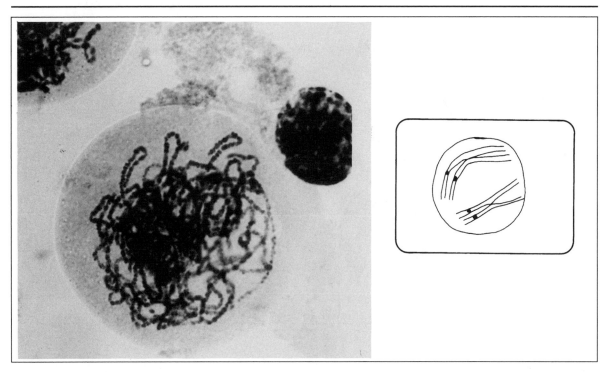

Figure 4.11 Pachytene in *Lilium* cv. Enchantment. Photograph courtesy of Professor J.S. Parker.

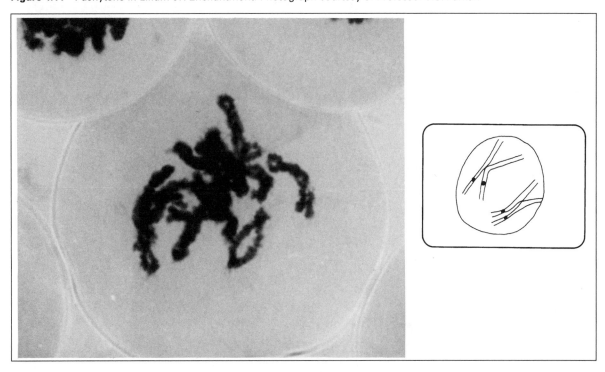

Figure 4.12 Diplotene in *Lilium* cv. Enchantment. Photograph courtesy of Professor J.S. Parker.

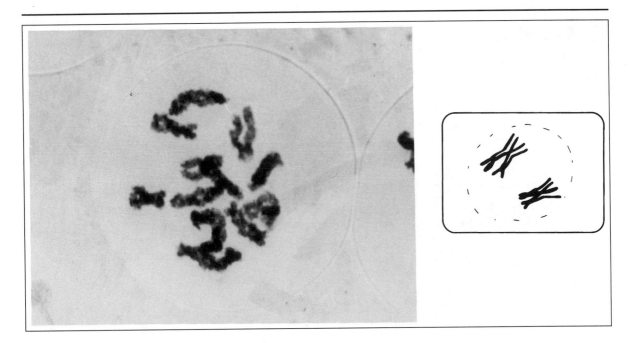

Figure 4.13 Diakinesis in *Lilium* cv. Enchantment. Photograph courtesy of Professor J.S. Parker.

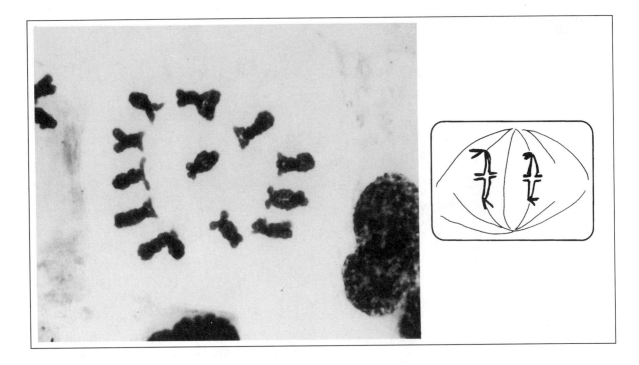

Figure 4.14 First metaphase in *Lilium* cv. Enchantment. Photograph courtesy of Professor J.S. Parker.

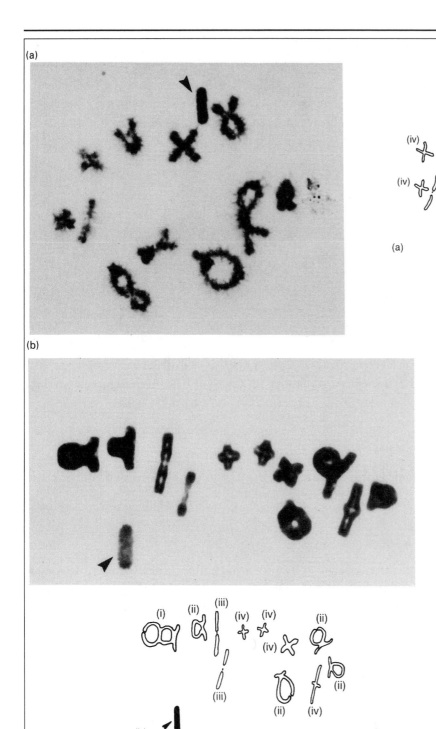

Figure 4.15 (a) Diakinesis and (b) metaphase I in *Schistocerca gregaria* (locust). The sex univalent is arrowed. Pairing configurations: (i) bivalent, three chiasmata; (ii) bivalent, two chiasmata, ring formed; (iii) bivalent, one terminal chiasmata; (iv) bivalent, cross-shaped, one chiasmata. Photograph courtesy of Professor J.S. Parker.

Diakinesis (Figure 4.13)

Contraction of the chromosomes is nearly maximal. The nuclear membrane dissociates. The paired chromosomes structures, held together by chiasmata, rotate in various planes so that they position themselves in a state of maximum repulsion and start to orientate on the metaphase plate.

Prometaphase

The bivalents attach to the newly formed spindle. The bivalents are now of characteristic shapes depending on the number and distribution of chiasmata.

First metaphase (Figure 4.14)

The chromosomes lie on the equatorial plate, centromeres attached to the spindle fibres.

The last two main stages described above (diakinesis and metaphase) can be somewhat indistinct in *Lilium*. Therefore examples of these stages in *Schistocerca gregaria* (locust) are shown in Figure 4.15.

First anaphase (Figure 4.16)

The bivalents separate and the homologues are pulled to opposite poles.

First telophase and interphase

This is often a very rapid process such that cytokinesis may not occur. There is no replication of DNA so each nucleus contains half-bivalents, i.e. the haploid chromosome number but the same number of chromatids as in mitosis.

4.3.2   Second meiotic division (Figure 4.17)

This takes place in the two cells formed in the previous set of divisions. The chromosomes enter this phase as chromatids joined by a centromere. They align on the metaphase plate of the newly formed spindle (metaphase II).

The centromeres split (as described in mitosis above) and one daughter chromatid moves to each pole (anaphase II, Figure 4.18).

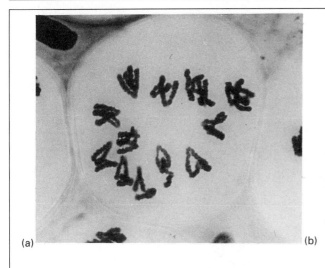
(a)

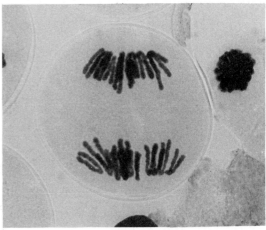

(b)

Figure 4.16 First anaphase in *Lilium* cv. Enchantment. (a) Early anaphase with poleward movement just starting. (b) Late anophase/telophase. Photographs courtesy of Professor J.S. Parker.

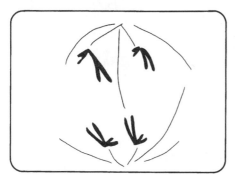

At telophase II (Figure 4.19), the interphase nuclei are reformed and cytokinesis occurs, forming four haploid daughter nuclei (Figure 4.19b).

A brief summary of some of the differences between mitosis and meiosis is given in Table 4.2.

Some of the stages of meiosis (pairing, synaptonemal complex and recombination) will now be discussed in more detail.

4.4 Chromosome pairing

Little is known about the mechanism of chromosome pairing (or synapsis). It has been proposed that homology is checked by direct comparison of DNA sequences between chromosomes. Initially two chromosomes come into close contact within the nucleus. A small

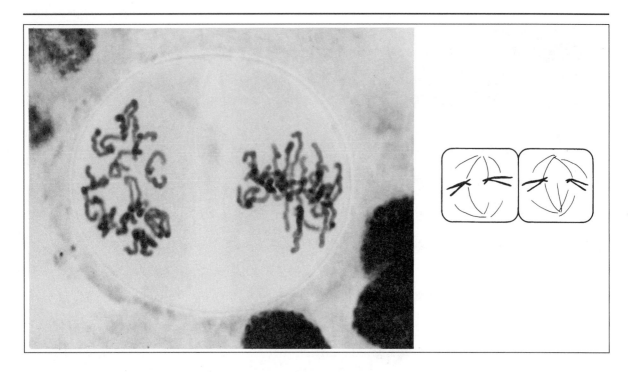

Figure 4.17   Second meiotic division in *Lilium* cv. Enchantment. Photograph courtesy of Professor J.S. Parker.

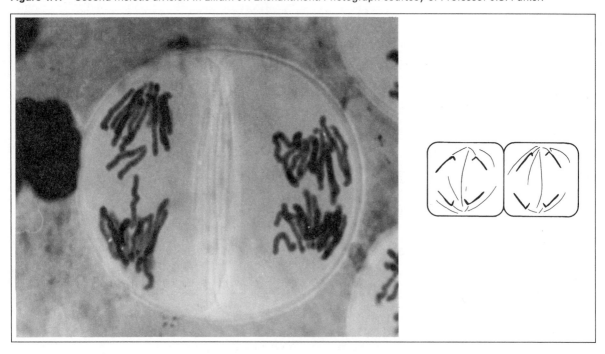

Figure 4.18   Anaphase II in *Lilium* cv. Enchantment. Photograph courtesy of Professor J.S. Parker.

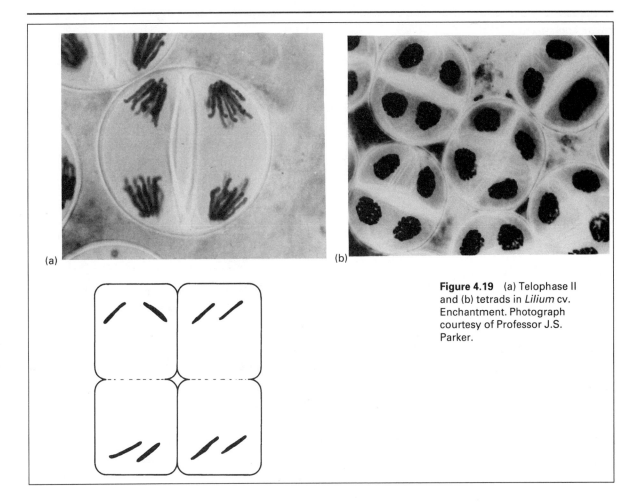

Figure 4.19 (a) Telophase II and (b) tetrads in *Lilium* cv. Enchantment. Photograph courtesy of Professor J.S. Parker.

segment of synaptonemal complex (SC) is formed between them followed by a check for extended sequence homology by strand invasion via either a D-loop or a single-stranded nick (Figure 4.20).

If extensive homology is found then the two homologues pair by a zippering effect of the continued SC formation. If no sequence

Table 4.2 Main differences between mitosis and meiosis

Mitosis	Meiosis
Nucleus divides once	Nucleus divides twice
Chromosomes divide once	Chromosomes divide twice
Two daughter nuclei	Four daughter nuclei
Constant chromosome number	Chromosome number halved
Chromosomes first seen double	Chromosomes single when first seen
Homologues independent	Homologues associate in pairs
Products genetically identical	Products genetically non-identical if the original cell is heterozygous

Figure 4.20 Homology checking via D-loop formation and single-stranded nicks.

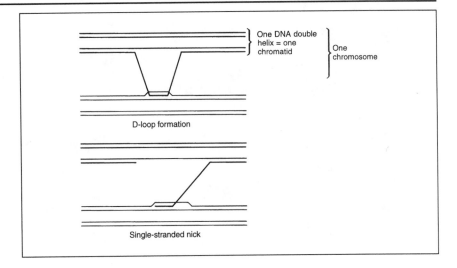

homology is found, then the SC dissociates and the two chromosomes are free to try another homology check elsewhere. The dilemma for the organism is 'how perfect must the sequence homology be to define a homologous chromosome?'. Do small allelic differences, measured in single base pair differences matter, or are they ignored? Outbreeding species have a higher degree of polymorphism than inbreeders and therefore must have a more flexible checking system. Ultimately, there must be a preprogrammed checking level built into the organism, but how this is defined or regulated is as yet unknown.

There is also another constraint, in that hybrid DNA formed by single-strand transfer is prevented from progressing into a full chiasma at this stage. This is important to prevent non-homologues taking part in genetic exchange and potentially causing large-scale chromosome aberrations. If the match is successful, the single-strand transfers may remain as potential sites for full chiasmata formation or they may lapse and undergo gene conversion (see section 4.7). Alternatively, all chiasmata may start afresh. This outcome may again depend on the individual species and may well reflect the overlying cellular mechanism for highly non-random distribution of chiasmata.

This model implies that pairing, while ultimately very specific, is initially a 'hit and miss' affair between all chromosomes within the nucleus.

Some organisms contain genes that prevent homoeologous pairing (i.e. pairing between non-homologues), e.g. wheat, oats, *Lolium*. The wheat *Ph1* gene on chromosome 5BL is particularly well known, although its mode of action is poorly understood. These genes are useful in polyploids, in which they prevent pairing between homoeologous chromosomes, which in simplistic terms have a diploidizing effect on the chromosomes. This can have a beneficial effect by enhancing pairing between homologues, reducing multivalent formation and potentially increasing fertility.

4.5   Synaptonemal complex formation

Synapsis (pairing) usually results in the formation of the synaptonemal complex (SC) (Figure 4.21). This is a ribbon-like structure, formed

Figure 4.21 The synaptonemal complex. (a) Basic structure. (b) Longitudinal section of SC in *Corthippus brunneus* showing both lateral and central elements. Courtesy of Dr C. Tease.

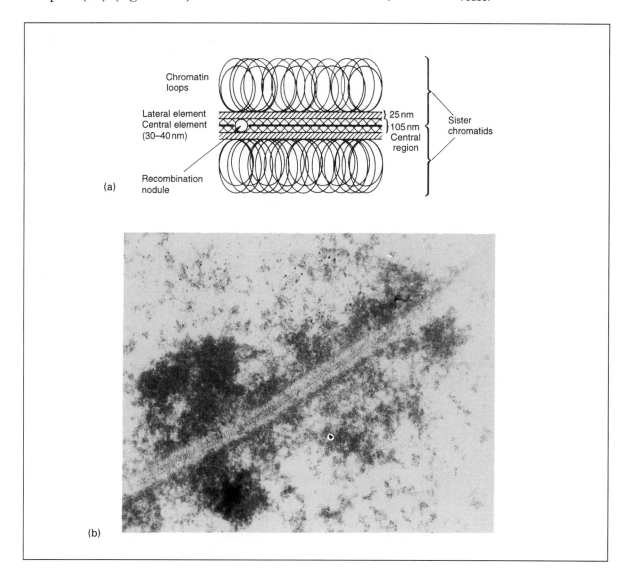

Figure 4.22 Photograph of SC from a whole cell of *Hypochoeris radicata*; centromeres are arrowed. Courtesy of Dr K. Hall.

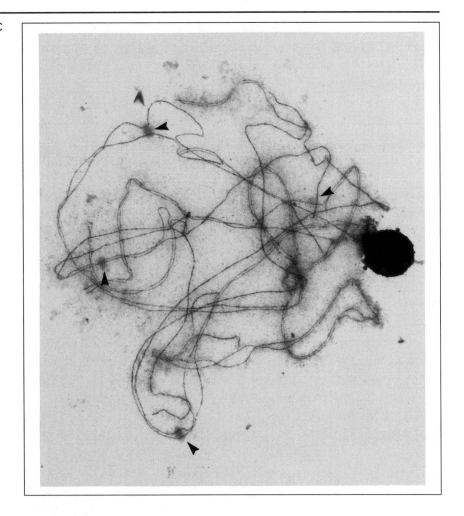

between homologous chromosomes, which can only be viewed under the electron microscope after special staining procedures (Figure 4.22). SCs are remarkably conserved between species, reflecting the evolutionary importance of their role. They consist of two electron-dense lateral elements (mean width 25 nm) separated by a less dense central region of approximately 105 nm containing a central element of 30–40 nm. The lateral elements consist of proteins, particularly the basic proteins such as arginine and lysine, and are derived from each pair of homologous chromosomes. Most of the chromatin is located outside the complex and is folded into loops that are attached at their base to the lateral element. The central element is also protein with small amounts of RNA. The major components are synthesized *de novo* during meiotic prophase and are not rearrangements of pre-existing compounds in the nucleus.

Proposed roles for the SC include the following.

1. They may constitute a structural framework within which recombination can occur. Experiments and the analysis of mutants indicate that the SC is important for effective recombination.

    *   Mutants defective in synapsis invariably do not undergo recombination.
    *   In organisms in which SC formation is limited in extent, recombination is similarly limited.
    *   Some organisms that do not undergo meiotic recombination (e.g. male *Drosophila*) do not form SCs.

    However, some organisms, such as desynaptic mutants in barley and tomato and haploid barley, have normal SCs but have virtually no chiasmata. Therefore, the presence of SCs is essential, but not wholly responsible, for crossing over at pachytene.

2. SCs may be important in stabilizing crossover sites and the regular disjunction of chromosomes. It is believed that the SC plays a role in stabilizing crossover sites while chromatin bridges form. In some organisms, after dissolution of most of the SC, some remnants remain associated with chiasmata. Two yeast mutants that cannot assemble the SC, but show greatly reduced levels of crossing over have been isolated. The crossovers that do occur do not ensure proper disjunction of the chromosomes in meiosis I and therefore are defective in some way.

3. SCs may play a role in the mechanism of interference of chiasmata. Chiasmata are non-randomly distributed along the length of the chromosome arm. They also 'interfere', such that if one chiasma is present it precludes the formation of another one within a certain distance. This is thought to result from the transmission of a signal originating at the site of crossing over and extending outwards in both directions to inhibit additional nearby crossovers. It is thought that the SC may somehow be involved in this information transfer.

4.6   Recombination nodules

There are other transient structures often associated with the SC at zygotene and pachytene, generally with the central elements. These are called recombination nodules (Fig 4.23).

There are two recognized classes of nodule which differ in appearance, frequency, distribution and timing.

Figure 4.23 Section of SC preparation showing central centromere with two recombination nodules either side. Reproduced from Albini and Jones (1984), *Experimental Cell Research*, **155**, 588–592, Figure 2, with permission of Academic Press.

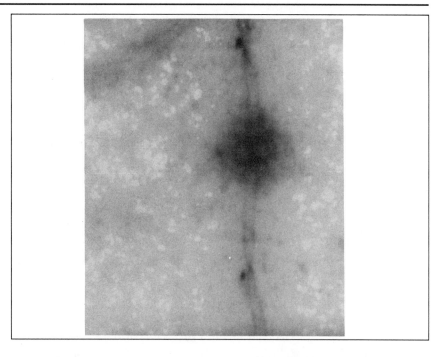

1. Early nodules may be ellipsoidal in shape in some organisms, and confined to zygotene when chromosome pairing is in progress. They are randomly distributed along the SC and are approximately twice as numerous as late nodules, although in tomato they are 30 times as numerous as late nodules. It has been suggested that they may be associated with homology checking or gene conversion events.

2. Spherical shaped late nodules are detected at late zygotene/ pachytene, when chromosome pairing is completed. They are present in much fewer numbers than the early nodules and, since they often mirror crossover events in terms of number and distribution, it has been assumed that they are associated with this process. It is believed that nodules mark the sites of multienzyme complexes that catalyse recombination, although how these operate and what determines the genes at which they are located is largely unknown. Autoradiographic experiments have shown that repair-type DNA synthesis occurs in the vicinity of late nodules and supports the idea that these somehow mediate reciprocal exchange and conversion events.

Synaptonemal complexes are routinely used to study pairing relationships between chromosomes. The SC readily reveals any misalignment or loop formation (Figure 4.24) due to deletions and duplications and

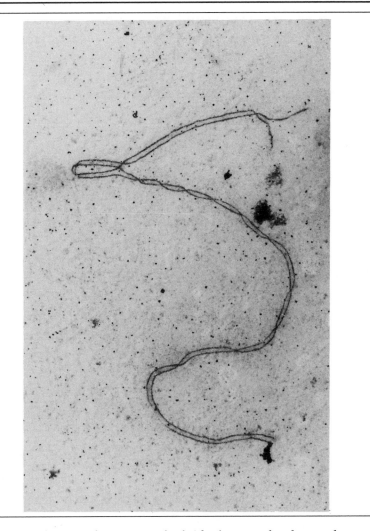

Figure 4.24 Section of SC showing an inversion loop between two pairing partners in *Hypochoeris radicata*. Courtesy of Dr K. Hall.

the extent of pairing between polyploids than can be detected at metaphase I at the light microscope level.

In addition to the study of pairing, SCs are also being used to karyotype organisms with very small chromosomes, particularly plants such as the brassicas. SCs under the EM present an elongated 'chromosome' which can be karyotyped according to arm ratios and length far more accurately than mitotic metaphase spreads.

4.7 Recombination and gene conversion

Labelling techniques have demonstrated that chiasmata correspond to sites of physical breakage and reunion between non-sister chromatids (commonly called recombination). The chiasmata can easily be viewed under the light microscope (at diakinesis and metaphase I), but they only represent reciprocal crossovers; non-reciprocals cannot be detected. It must be borne in mind that they are indicative of an event which, in molecular terms, only encompasses a few hundred to a few thousand base pairs.

The basic model is that of Holliday, which is described here (Figure 4.25). The model works on the assumption that each chromatid is a

Figure 4.25 Diagram of recombination process, after Holliday.

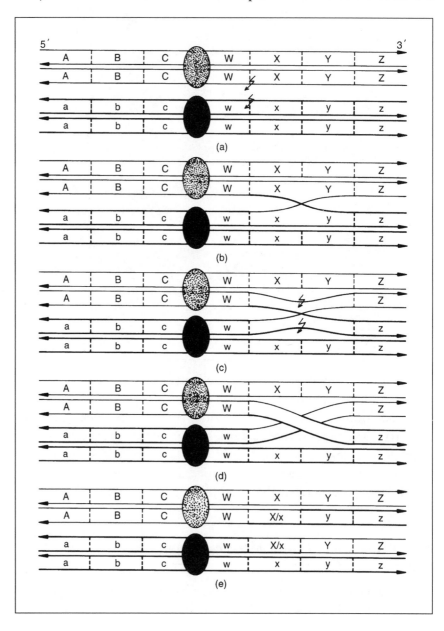

single double helix of DNA and that recombination takes place after chromosome pairing and replication.

1. Homologous chromosomes pair. Single-strand breaks occur in the chromatids of homologous chromosomes at identical sites in each (Figure 4.25a).
2. Strand separation and unwinding take place. The two separated strands are free to pair and anneal with the complementary chains of the homologues (Figure 4.25b). This results in the formation of a single-stranded bridge or half-chiasma.
3. The opposite strands of the chromatids break in identical places (Figure 4.25c).
4. The DNA unwinds and separates, the two strands anneal and a full chiasma is formed (Figures 4.25d and e). This cross shape between the two homologues is known as a cross-strand exchange or Holliday structure and has been seen under the EM *in vitro* and *in vivo*. This cross is cut by endonucleases, separating the two homologues, producing recombination of outside markers if the homologues contain different information. This would not hold true for pure inbred lines.

There is a potential region of hybrid DNA (Figure 4.25e) which, if mismatched base pairs are present, is potentially unstable and may be repaired by the correction enzyme systems. If repair is not effected aberrant segregation ratios such as 5:3 can occur.

Not all half-chiasmata continue through the process to full recombination. If the half-chiasmata can break, the nicks in each chromatid are repaired and any potential mismatches may be resolved by repair mechanisms. This results in non-mendelian gene segregation and is called gene conversion.

At first glance, it may appear that there is an equal chance of gene conversion and full chiasmata formation, however the process is heavily loaded towards recombination.

Because the structures which arise in recombination are similar to those intermediates viewed in DNA damage and repair, it is thought that perhaps the same or similar sets of genes are functional in each process.

This basic theory has been elaborated on since, to accommodate other observations in meiosis. These include the 'Aviemore model' proposed by Meselson and Radding, which is biased towards asymmetric events, the RecA model put forward by Radding, the 'double-strand break repair model' suggested by Szostak and the RecBC model proposed by Smith. However, it is not known as yet, which, if any, is 100% correct, and it is suspected that different mechanisms may operate under different circumstances. However, the model of Holliday is probably the simplest to understand and provides a basis on which to examine and understand the other more complex models.

4.8   Meiosis in polyploids and structural rearrangements

The division processes described above were all based on a diploid chromosome complement. However, this is frequently not the case. Plants, in particular, are often polyploid, and chromosome translocations and inversions may be present in the organism's genome. These may cause problems not only of altered dosage of genes, but also, during regular disjunction in meiosis, in maintaining a balanced chromosome set to pass onto the offspring.

4.8.1   Polyploids

A polyploid individual has more than two complete sets of chromosomes, of which there are two types:

1.   autopolyploids, in which all sets originate from the same species;
2.   allopolyploids, in which the extra sets (which may or may not be homologous) originate from different species.

Because of these multiple sets, segregation problems may occur, such that the gametes may not receive full sets of chromosomes and, as a consequence, the organism will suffer reduced fertility.

For example, autotriploids ($2n=3x$) have three complete chromosome sets. At any one point only two chromosomes can pair effectively (although small sections of triple-paired synaptonemal complex have been seen in some triploids and trisomics). This means that combinations can occur of three univalents, one univalent and a bivalent or a trivalent. Univalents, as discussed below, will generally be lost from the cell via micronucleation. Bivalents will segregate normally. The orientation of the trivalent on the metaphase plate determines how many chromosomes go to each pole (Figure 4.26).

To ensure full fertility, all chromosomes must form trivalents, all trivalents must orientate in a convergent manner and all convergent trivalents must segregate two chromosomes to one pole and one to

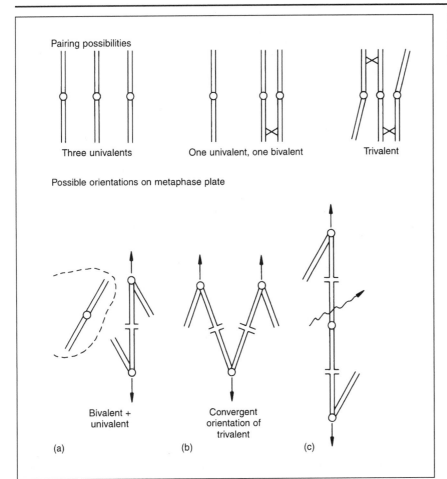

Figure 4.26 Some pairing possibilities in a triploid with orientation on the metaphase plate.

the other. This would produce two diploid gametes and two haploid gametes. Needless to say, this rarely happens and triploids form all types of configurations at metaphase I. This produces unbalanced gametes that are generally non-viable, and consequently fertility is reduced (Figure 4.27).

Table 4.3 Frequencies (%) of lagging chromosomes at anaphase I (chromosomes that show retarded or no movement compared with the others) in autotriploids.

Species	No. of laggards per cell							Total no. of cells
	0	1	2	3	4	5	6	
Lolium perenne	37.0	26.0	17.0	9.0	9.0	1.0	1.0	1103
Hordeum spontaneum	29.92	26.26	21.26	15.35	2.36	3.15	0.79	127
Triticum monococcum	89.20	7.2	1.20	2.4	0.0	0.0	0.0	250

Data from Singh (1993) with permission.

Figure 4.27 Example of pairing configurations at metaphase I of a triploid, *Hypochoeris radicata*. Configurations from left to right: frying pan trivalent; linear trivalent; bivalent; trivalent in convergent orientation; univalent. Courtesy of Professor J.S. Parker.

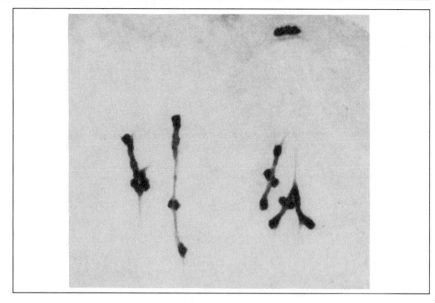

Examples of meiotic problems associated with triploids are shown in Table 4.3.

As a general rule, the higher the basic number, the lower the fertility. Other polyploids with odd numbers of chromosome sets – 5X, 7X, etc. – experience similar problems to triploids because the odd set, which will either form univalents, trivalents or complex multivalents. For example in a pentaploid when $2n=5x$ the chromosomes can pair in any combinations of between one and five chromosomes.

When a triploid has a total chromosome number of 9, fertility is reduced to 25% of the diploid level as a result of the probability of all chromosomes forming trivalents and all trivalents orientating the same way. When $2n=3x=12$, fertility is further reduced to 12.5% compared with the diploid. Higher basic numbers reduce fertility further and are almost completely sterile. In triploid apples such as Blenheim Orange and Bramley's Seedling, in which $2n=3x=51$, seed set is less than 5% and the viability of the seed set is well below 100%.

Even-numbered polyploids such as tetraploids ($2n=4x$), in general, have a higher level of fertility. Tetraploid chickpea has only a 15% reduction in fertility when compared with the diploid.

Tetraploids have four complete sets of chromosomes. Each set of four homologues has the following pairing options: four univalents; two bivalents; a bivalent and two univalents; a univalent and a trivalent; or a quadrivalent. The frequency with which these are produced depends on the pairing arrangements. The same basic rule as described for triploids also applies here: segregation of complete sets of chromosomes produces

Species	Univalents	Bivalents	Trivalents	Quadrivalents
Arrhenatherum elatius	0.0	4.3	0.0	4.8
Avena strigosa	0.1	5.0	0.1	4.4
Hordeum bulbosum	0.1	5.6	0.2	4.0
Hordeum vulgare	0.3	5.7	0.1	3.9
Pennisetum americanum*	2.64	8.97	0.38	1.49
Petunia hybrida	0.1	4.6	0.1	4.6
Pisum sativum	0.3	5.21	0.16	4.2
Secale cereale	0.5	6.1	0.2	3.7
Triticum monococcum	0.2	3.5	0.1	5.1
	0.62	9.86	0.23	1.74

Data from Singh (1993) with permission.
*Pennisetum glaucum.

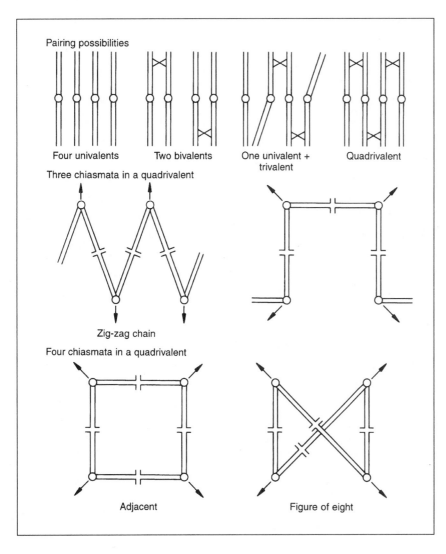

Pairing possibilities

Four univalents   Two bivalents   One univalent + trivalent   Quadrivalent

Three chiasmata in a quadrivalent

Zig-zag chain

Four chiasmata in a quadrivalent

Adjacent   Figure of eight

Table 4.4 Chromosome configurations observed in microsporocytes at metaphase I in tetraploid species with $2n = 4x = 28$

Figure 4.28 Some pairing possibilities in a tetraploid with orientation on the metaphase plate. All examples shown produce balanced gametes.

Figure 4.29 Example of pairing configurations at metaphase I of a tetraploid, *Hypochoeris radicata.* Configurations from left to right: quadrivalent adjacent; quadrivalent alternate, quadrivalent indifferent, bivalent; bivalent. Courtesy of Professor J.S. Parker.

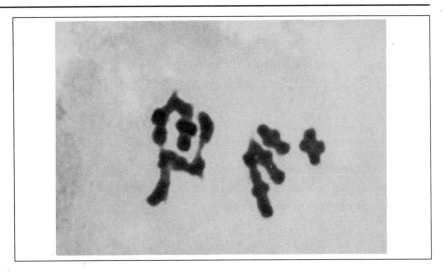

viable gametes; other combinations either produce non-viable gametes, or result in inviable or deleteriously affected offspring.

Examples of meiotic configurations in several autotetraploids are given in Table 4.4.

Many tetraploids have mostly bivalent and quadrivalent pairing as this allows for maximal pairing in the synaptonemal complex, producing balanced segregation at anaphase I (Figures 4.28 and 4.29).

Many even-numbered polyploids may be allopolyploids with little similarity between the different contributing chromosome sets. Thus, in order to maintain fertility, bivalents are only formed between homologous chromosomes, for example the hybrid formed between *Brassica oleracea* ($2n=2x=18$) and *Raphanus sativus* ($2n=2x=18$). The F1 is sterile owing to lack of pairing between the two chromosome sets, but if these are doubled the allotetraploid ($2n=4x=36$) is fertile with the formation of 18 bivalents.

Bivalents may also be formed in even-numbered polyploids owing to the effects of various pairing control mechanisms (such as *Ph* in wheat) which suppress homoeologous pairing.

The addition of B chromosomes to a polyploid cell can also increase the rate of bivalent formation. For example, autotetraploid rye with two Bs shows an increase in bivalent formation of approximately 15% when compared with the same variety without Bs.

4.8.2　Univalents

If homologues fail to pair or no chiasmata are formed between them, univalents result at metaphase I.

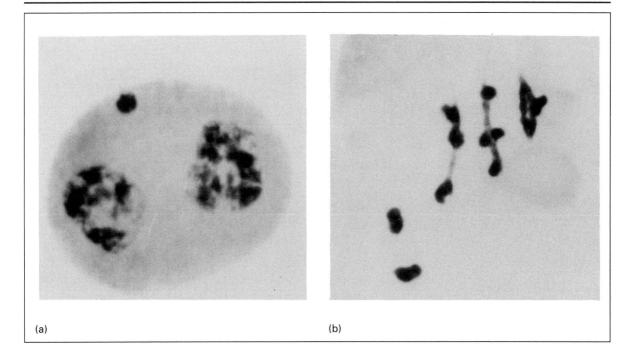

(a)　　　　　　　　　　　(b)

Their subsequent behaviour at anaphase I is unpredictable and variable. They may not be able to move to the equator in time for bivalent separation and may lag around the periphery of the cell. If they do not engage in successful division, they may be lost in the form of micronuclei. Sometimes they may congregate on the equator and divide at the centromere at anaphase I to form two chromatids. At the second division they will not be able to divide correctly and are, as a result, lost as micronuclei (Figure 4.30).

Another potential fate is misdivision (Figure 4.31). This occurs when the centromere splits transversely, i.e. between the two arms, to form two telocentric chromosomes, which can then undergo division. Many of these are unstable, and the sticky ends from identical arms may fuse to form an isochromosome (i.e. deletion of one whole arm and duplication of the other), the consequence of which is genic imbalance.

Figure 4.30 Examples of univalent behaviour. (a) Micronucleus present in cell. (b) Metaphase I of *Hypochoeris radicata*. Configurations from left to right: ring bivalent; rod bivalent; rod bivalent; two univalents (not orientated on metaphase plate). Courtesy of Professor J.S. Parker.

4.9 Structural changes

Structural changes such as inversions and reciprocal translocations may also affect the outcome of meiosis as they may reduce the possibility of forming a balanced set of gametes.

Figure 4.31   Misdivision of a
univalent.

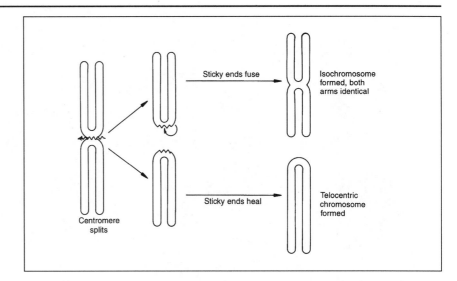

Figure 4.32   (a) Paracentric
inversion. (b) Pericentric
inversion.

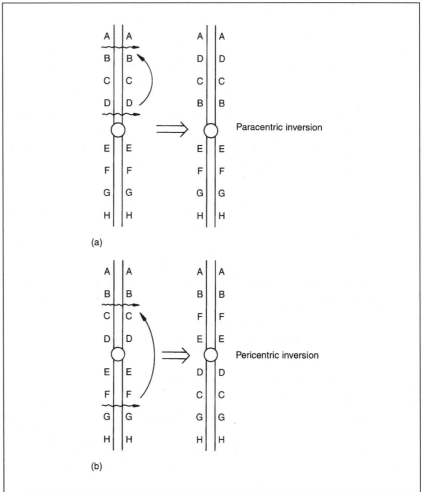

4.9.1 Inversions

These are formed when there are two breaks in a single chromosome and the segment between the two breakpoints realigns and rejoins in the opposite orientation. The breakpoints can be on opposite sides of the centromere in acro- or metacentric chromosomes (pericentric inversion, Figure 4.32a) or take place in one arm (paracentric inversion, Figure 4.32b).

The central segment is now in opposite polarity to its homologue in a heterozygous individual. To enable effective pairing in meiosis, an inversion loop is formed, the size of which is dependent upon the size of the segment involved.

With paracentric inversions, if no chiasmata are formed in the reverse loop, but occur elsewhere on the bivalent, segregation occurs normally. However, if a single chiasma forms within the inversion loop the two

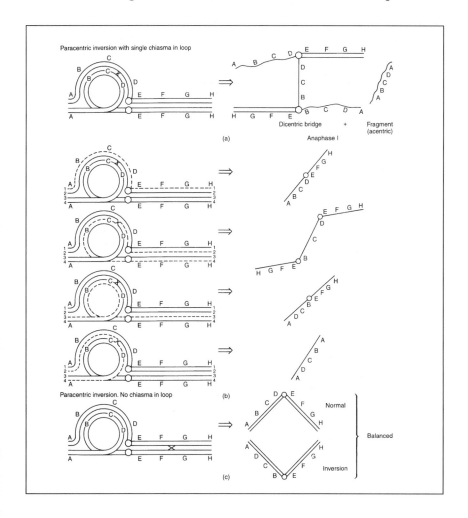

Figure 4.33 (a) Demonstration of the result of one chiasma forming in the inversion loop of a paracentric inversion. (b) Explanation of how you work out the results. By numbering the chromatids at each end, start with (1) and follow it through. Continue until all at one end are used up and then check that all opposite ends have been accounted for as not all results include a centromere. (c) Result of no chiasma formed in a paracentric inversion loop.

chromatids involved in the chiasma will form a dicentric bridge and acentric fragment at anaphase I (Figure 4.33).

The fragment will be lost (as it has no centromere and thus cannot interact with the spindle apparatus) and the bridge will come under strain and may break at a random position, the final result being gametes with duplications and deletions that may not produce viable offspring. The two chromatids not involved in the chiasma separate normally to produce viable gametes. There will be two out of four viable products with one exchange in the loop.

In the pericentric inversion, bridges and fragments are not formed by chiasmata within the reverse loop, but gametes formed from the recombination products will be deficient in some gene sequences and have others duplicated with the consequent problem of unbalanced gametes (Figure 4.34).

Owing to the deleterious effect of chiasma and recombination within an inversion loop in both para- and pericentric inversions, the overall effect may be reduced transmission of the recombinant products.

In humans paracentric inversion carriers are clinically normal, but reproduction is affected. If a crossover occurs within the inversion loop, generally the degree of imbalance produced is incompatible with viability in all but exceptional cases. This is not true of pericentric inversion carriers, who are also clinically normal. They risk producing an affected offspring, the chance of which depends on the origin of the inversion. If it originated with the father there is an approximate 4% chance of

Figure 4.34 Results of pericentric inversion with (a) a single chiasma in loop. Unbalanced gametes formed, and (b) no chiasma in loop. Balanced gametes formed.

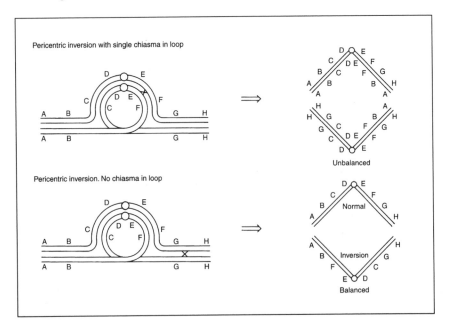

producing an abnormal child. If the inversion originated with the mother, the risk is increased to 8%.

Examples in plants show increased infertility, reflected in the measurable levels of pollen abortion. For example, in the series of paracentric inversion in maize, In2a has a pollen abortion level of between 12.4% and 21.6%, In3a 11.5–27.6% and In4a 28.2%. The normal diploid level is around 2.9%.

Pericentric inversions have more serious effects. Two heterozygous pericentric inversions in bean (*Vicia faba*) had pollen abortion levels of 50%. Returning to examples in maize, pericentric inversion Inv2b had average levels of 19.1% and 20.1% and Inv5a average levels of 28.3% and 12.5% of pollen and ovule abortion respectively. The figures are higher in the male owing to a higher frequency of crossing over in male flowers.

The fact that these types of rearrangements persist in populations, particularly in sex chromosomes, has been proposed as a possible mechanism for preserving advantageous combinations of gene blocks and families.

4.9.2   Reciprocal interchanges

This is the result of a break in two non-homologous chromosomes followed by the reunion of particular broken ends of different chromosomes. This is a reciprocal exchange and involves no loss of material (Figure 4.35).

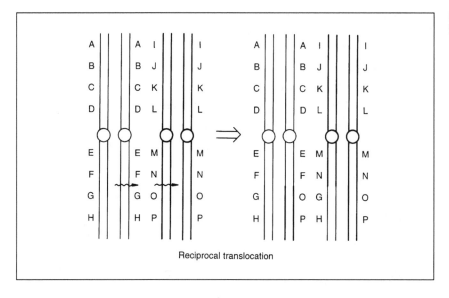

Reciprocal translocation

Figure 4.35 Reciprocal translocation.

Figure 4.36 Pairing configuration of reciprocal translocation at pachytene, such that all segments pair.

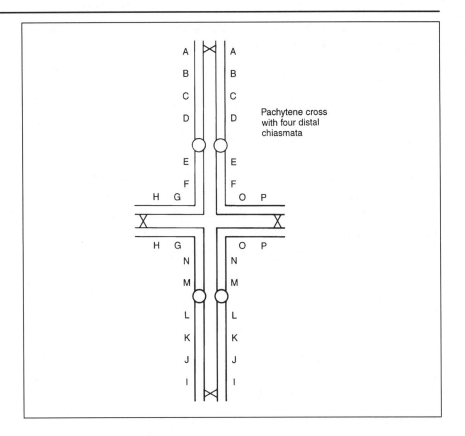

Pachytene cross with four distal chiasmata

Complete pairing is produced by association of all four chromosomes, the two involved in the interchange and the two standard ones in the form of a pachytene cross (Figure 4.36).

If one chiasma is formed at the end of each chromosome a quadrivalent will be formed. The orientation on the metaphase plate will be either an adjacent or figure-of-eight configuration, with only the latter producing genically balanced gametes (Figure 4.37). If fewer chiasmata are formed, a chain of four may result or trivalents and a univalent.

In humans the birth frequency of balanced translocations is approximately 1 in 250. Problems occur in meiosis and result in the production of unbalanced offspring. The actual risk of this occurring is about 12%. The calculated risk is much higher, but gametic selection operates to a certain extent and many abnormal fetuses spontaneously abort.

Other chromosomal abnormalities such as deletions, duplications and aneuploids have not been discussed here, but examples will be given in later chapters. The first two are mainly effective through their alteration in gene dosages and in general do not disrupt meiosis to the extent described in the examples here.

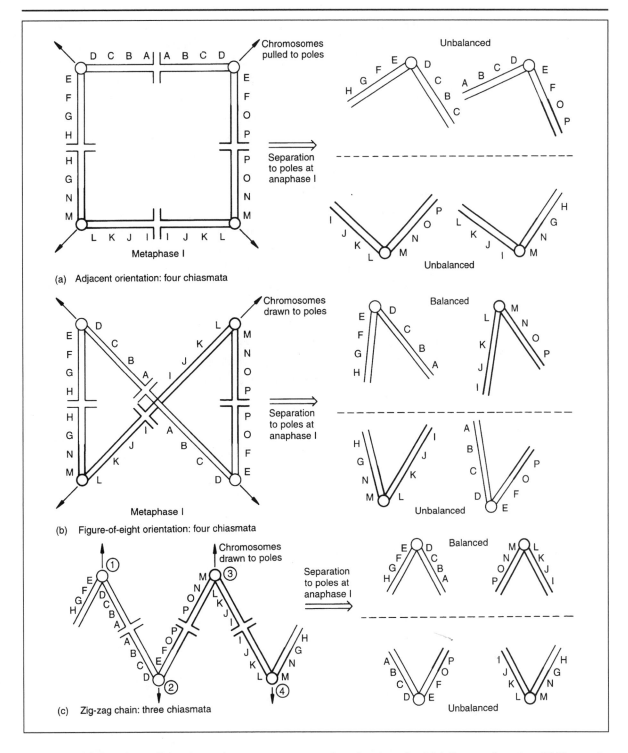

Figure 4.37   Orientations of interchange heterozygote at metaphase I and results. (a) Adjacent orientation. (b) Figure of eight. (c) Zig-zag chain.

Aneuploids result in some infertility owing to abnormal segregation of the extra chromosome(s), which are frequently lost at meiosis as univalents. The aneuploids themselves may be morphologically distinct. In plants they are often produced and maintained to aid in the production of chromosome substitution lines and gene mapping experiments, but are not commercially useful because of chromosomal instability. As regards pairing possibilities, they represent a variant of polyploids. The disruptive effect of the extra chromosome(s) on meiosis and the subsequent ability to mature into a viable organism is very species and chromosome dependent. The classic example of this is with man, with only three types of autosomal trisomic fetuses (chromosomes 13, 18, 21) progressing to full term and birth; all others are spontaneously aborted *in utero*. All aneuploid plants show reduced fertility and segregation for the extra chromosome in the offspring. Maintenance of such stocks for mapping purposes requires continual screening and selection procedures.

The process of nuclear division underpins the successful maintenance and reproduction of an organism. Errors lead to disease and infertility. Understanding these processes, particularly at a molecular level, will provide many answers to genetic disease, for example cancer cell proliferation, and lead to the enhanced capacity for genome manipulation. Much of the molecular work has been carried out on yeast; it is still largely incomplete, and the extent to which the results apply to higher organisms is unknown and for this reason will not be discussed here.

Gene control by position and origin

5

Summary

This chapter describes two phenomena that are not obviously related: chromosome organization within the nucleus and imprinting.

Chromosome organization within the nucleus encompasses the hierarchical position of a chromosome within the nucleus, which can affect its stability and consequent expression of genes. Chromosomal imprinting describes the maternal or paternal origin of a chromosome or gene that determines its expression.

These topics have been combined in this chapter because they have only recently been subjected to molecular investigation and hence little is known about them. However, it is becoming clear that their effects are widespread and dramatic, and the true extent of these effects is only beginning to be understood. They also demonstrate that gene expression is subject to many more forces than originally supposed and is not solely directed by gene sequence; a superimposed higher order of control exists.

5.1 Chromosome organization within the nucleus

This section will examine the evidence for superimposed chromosomal positioning within the nucleus and nuclear compartmentalization.

Chromosome positioning within a nucleus is essentially a dynamic affair resulting from transcription, replication, division processes, etc. When looked at under the microscope, the interphase nucleus appears to resemble a bowl of spaghetti with no distinct order. However, recent studies indicate that there may well be a superimposed pattern. As long ago as 1885, Rabl noted that mitotic metaphase chromosomes align within a nucleus with their centromeres facing one end of the cell and the telomeres facing the other: the so-called 'Rabl configuration.' Since then geneticists have debated whether chromosomes do indeed occupy distinct individual domains within a nucleus and whether it is possible to relate nuclear architecture to aspects of gene expression and chromosome behaviour.

Nuclei exhibit very different, cell type-specific shapes that depend on the relative condensation and position of the chromosomes they contain. However, their shape is not fixed, and can rapidly change in response to the environment, for example when the cells are placed in tissue culture. Why do these changes occur and what effect does a change in chromosome condensation have on nuclear activity?

The need for organization of chromosomes during meiosis is self-evident: without strict control of distribution, associations of inter-locked bivalents and lethal non-disjunctions would inevitably occur.

The importance of the distributions and associations that are seen during studies of mitosis is rather more difficult to explain for the simple reason that such associations may not reflect the true state of chromosome distribution in life. Much work on mitosis is carried out using squashes of metaphase chromosomes, which is obviously an artificial situation. In addition, when cell division is arrested at metaphase by spindle poisons the position of chromosomes tends also to be disrupted. However, it is possible to draw general conclusions from studies of large numbers of metaphase plates. It would seem likely that under such circumstances associations which occur regularly are a reflection of the state of the cell in life.

Although evidence for an ordered structure has been found in species with a small number of chromosomes, such as *Crepis capillaris* ($2n=2x=6$), in organisms such as humans, in which the number of chromosomes per cell is much higher, the distribution of chromosomes during metaphase is much more difficult to define. Observation of numerous human metaphase spreads has yielded general indications of chromosome disposition. The Y chromosome tends to be peripheral. Telocentric chromosomes can be associated by their satellite ends. These particular chromosomes all have the common function of organizing the nucleolus and so might be expected to be found together.

An intensive study using metaphase spreads to examine chromosome organization has been carried out on plants. *Hordeum vulgare* cv.

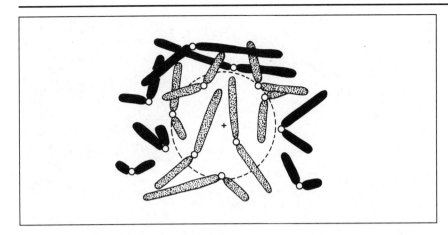

Figure 5.1 Reconstructed metaphase plate in a root tip cell of *Hordeum vulgare* cv. Sultan (stippled chromosomes) × *Secale africanum* (solid chromosomes) showing spatial isolation of the two chromosome sets. Reproduced from Bennett, M.D. (1988), *Kew Chromosome Conference III*, 198, Figure 2. Crown copyright. Reproduced with the permission of the Controller of Her Majesty's Stationery Office.

Sultan (barley) was crossed with *Secale africanum* (wild rye) and the F1 hybrids examined. The seven chromosomes of *Secale* could be clearly distinguished from the smaller *Hordeum* chromosomes. By analysing metaphase squashes and plotting the position of centromeres, it was possible to show the separation of the parental genomes, often in the form of concentric circles, as shown in Figure 5.1.

Further investigation showed that this was not an unusual finding, but occurred in roots of different ages, in anthers and in endosperm nuclei, i.e. the imposed pattern can persist through time and the development of organs and tissues.

Chromosome organization is thought to be under genetic control rather than the result of packing constraints, as in hybrids the organization of the two parental genomes is dependent not on their relative sizes but on the hybrid environment. For example, in the *Hordeum vulgare* × *Secale africanum* cross described above, the smaller *Hordeum* chromosomes lie in the centre of the nucleus, but in a similar hybrid, *Hordeum chilense* × *Secale africanum*, the *Hordeum* chromosomes, which again are smaller than those of *Secale*, are this time positioned around the outside of the cell. This finding has been verified in many other hybrids, i.e. the size of the chromosomes does not predispose them to a particular location (Figure 5.2).

A hybrid often resembles one parent more than the other, and it has been found that the resemblance is greatest to the parent that contributes the outer genome. Also, in unstable hybrids, chromosome loss occurs only in the genome with a peripheral location.

These findings in controlled crosses have been substantiated by studies on the somatic hybrid products of protoplasted tobacco cells. Digestion and fusing of protoplasts is much harsher on nuclei, but

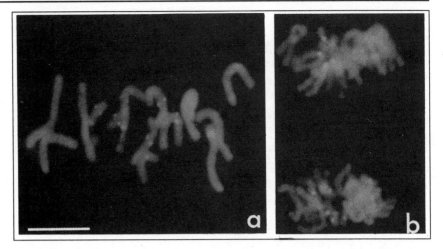

Figure 5.2 Genomic *in situ* hybridization enables the parental origin of the chromosomes to be determined in a sexual hybrid between barley × *S. africanum* (wild rye). The seven chromosomes of *S. africanum* origin are labelled and fluoresce yellow while the seven chromosomes from barley fluoresce orange with the propidium iodide counterstain. The two parental genomes lie in spatially separated domains at metaphase (a) and other stages of the cell cycle. At anaphase (b) all chromosomes have moved to the spindle poles. Reproduced with permission from Sumner and Chandley (1993), *Chromosomes Today Volume 11*, Chapman & Hall, figure originally from Leitch *et al.* (1991). Bar 10 µm.

superimposed patterning still persists. Hence, the positioning of chromosomes within a hybrid nucleus can affect their expression.

Recent advances in fields such as *in situ* hybridization, fluorescence microscopy and optical sectioning have enabled very accurate and far more realistic nuclear pictures to be built up. Individual chromosomes or specific chromosomal regions can be stained while the nucleus is still intact three-dimensionally. The position of each can then be analysed by optical sectioning and computers used to build up three-dimensional reconstructions of the whole structure.

While hybrids show that a superimposed organization exists throughout a variety of different hybrids, at a more refined level experiments have shown specific chromosome regions to be organized in a similar manner in several different species. This can be demonstrated using silver staining for active nucleolar organizing regions (NORs). In the human CNS (central nervous system) large neurons, cerebellar Purkinje cells, show larger and more numerous centrally placed NORs. Granule neurons in the cerebellum have only a few smaller and more peripherally placed NORs. This pattern exists in different species (mouse, hamster, guinea pig) of the same cell type, even though the NORs are located on different chromosomes in each species.

In a follow-up study, *in situ* hybridization was applied to the centromeres (alphoid repeats) of CNS cells. These repeats displayed orderly and reproducible cell type-specific patterns of organization. In both human and mice CNS cells repeats in distinct populations were organized in similar ways despite the fact that repeat sequences found in the centromeres of humans and mice differ considerably.

These studies together indicate that a common mode of cell-specific recognition and organization exists. This may either define or reflect functional differentiation at the level of nuclear chromosome organization. Interestingly, tumour cells, which are notorious for their chaotic organization, have random NOR and centromeric locations.

UV microirradiation to parts of the nucleus has been shown to damage discrete chromosome territories. These can now be defined using *in situ* hybridization of species-specific repeats and also whole chromosome probes (chromosome painting). The former was used on a human–mouse cell line containing a single human chromosome arm (4p). This displayed a reproducible and characteristic site in the nucleus of all cells studied. Another cell line with four human chromosomes (1,3,4,X) was examined using optical sectioning and it was found that each chromosome was associated with the nuclear membranes and was generally separate from the others and in a localized domain. How might these domains affect gene activity?

More detailed studies into chromosome domains within different cells of the CNS have been carried out. The position of chromosomes 1, 9, X and Y within the seizure foci of epileptic human brain cells was investigated by *in situ* hybridization using chromosome painting of single chromosomes. Standard positions were measured in 'normal' human brain cells and then compared with those from epilepsy sufferers. Dramatic repositioning of the X chromosome was found in epileptic foci. These changes were limited only to epileptic foci, not surrounding cells, and could not be explained simply in terms of altered pattern due to seizure activity. The implications of these studies are that such nuclear changes could underlie or give rise to this abnormal neuronal activity and that the position of a chromosome may affect its transcriptional activity. It is proposed that this repatterning is more or less permanent, established by, for example, trauma, developmental abnormalities or perhaps even the seizure activity itself.

Overall, studies have demonstrated reproducible cell type-specific chromosome domains in interphase nuclei (Figure 5.3). The famed Rabl configuration has only ever been demonstrated at metaphase; rapid dispersion occurs at anaphase.

How might these domains be organized? It has been suggested that centromeric and other highly repeated sequences may act as general organizing centres for cell type-specific interphase patterns and that these are conserved in mammalian evolution. Such centres would allow chromosome arms to extend into and contract from an interior transcriptionally active nuclear compartment. Such a division of nuclear space has been demonstrated by localization of nuclear poly-A RNA. It has been shown that transcription domains with defined borders often

Figure 5.3   Reconstruction of a nucleus showing nuclear envelope (solid lines), nucleolus (broken lines) and centrometric structures (solid areas). Note that all the centromeres are clustered near the surface of the nucleus. Reproduced from Heslop-Harrison *et al.* (1988), *Kew Chromosome Conference Conference III*, 214, Figure 2.

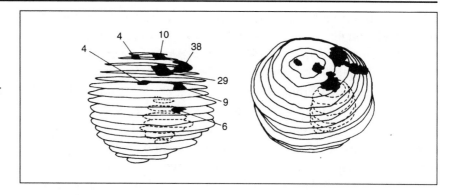

surround nucleoli in an interior location in areas of very low DNA concentration. They are clearly separate from areas of late-replicating DNA, i.e. there is distinct physical partitioning of two separate nuclear functions during late S-phase. Studies on the localization of antigens associated with transcriptional enzymes also show a defined central location in interphase nuclei.

How might this nuclear architecture be maintained? Experiments by Gleba found that addition of spindle poisons to a cell destroys chromosome domains. These are restored later, on removal of the poison. Hence, microtubules are implicated in maintenance of nuclear architecture. Proteins which are either implicated or for which there is direct evidence for a role in the movement of chromosomes are dynein and the group of microtubule motors generically called kinesin. Whether these are directly involved in interphase domains has yet to be clarified. Most of the work carried out on them so far has concentrated on their function during mitosis and meiosis.

Other candidates include a novel nuclear antigen AF-2. This is not directly associated with DNA replication but has been shown to be related to cell cycle alterations in chromatin structure. Also, a set of related proteins (PIKAs) has been localized to a large structural component in the nucleus of various cell types, the morphology of which changes dramatically during the cell cycle but has shown no correlation to either DNA replication or mRNA-processing activity.

So initial evidence indicates a potential superimposed pattern on chromosome organization within the nucleus. Perhaps this is not surprising in view of the fact that independent chromosome domains are attached by specific DNA sequences to a protein-based scaffold in a very ordered manner. Nuclear patterning is just one level further up from this. Several candidate proteins have been identified which may play a major role in this process. However, all of these investigations are at the preliminary level, but are increasingly turning up interesting results with implications for gene expression and disease causation.

5.2 Chromosomal imprinting

Chromosomal imprinting is a new area of investigation, and one of the most unpredictable. Imprinting, as the word implies, refers to the marking of a chromosome, either in part or as a whole, which renders the origin recognizable. The DNA has been marked for future use, the net result of which is that chromosomes can appear to contravene principles of mendelian genetics.

Evidence for imprinting first arose from studies of dipteran flies, in which eradication of entire sets of chromosomes is related to their parental origin. Other evidence comes from rare cases of uniparental disomy in humans. These show that it is not only the gene content of a chromosome that is important, but also which parent donated the sequence.

The discovery of imprinting in mammals was brought about in part by the ability to trace specific sequences, and consequently individual chromosomes, through generations. Although the precise mechanism of imprinting is as yet unknown, there are several hypotheses as to why this intriguing phenomenon occurs.

5.2.1 Functional description of imprinting

A chromosome, or part of a chromosome, that has passed through a maternal germline and is expressed differently when from one that has passed through a paternal germline can be said to be imprinted. These differences in expression may manifest themselves transitorily during development or have profound affects later on if the normal pattern of imprinting is disrupted.

The imprinted gene or chromosome is that one of a homologous pair that is non-functional. It should be remembered that this phenomenon is quite different from the psychological process of imprinting of newborns on parents.

5.2.2 Evidence for the phenomenon of imprinting

The first reports of imprinting were made during the 1960s and involved a dipteran fly, *Sciara coprophila*. This most unusual organism starts out as a zygote with two sets of autosomes (A maternal, A paternal) and three sex chromosomes (Xm, Xp, Xp), one maternal and two paternal. During embryogenesis one paternal X chromosome is lost in females so that the designation is now AmApXmXp. In males the position is slightly different in that both paternal X chromosomes are lost from the somatic tissues (AmApXm). During spermatogenesis the paternal autosomes are also lost at the first meiotic division (AmXm). Since the chromosomes which are lost are consistent in their origins it would be reasonable to assume that they are in some way marked, or imprinted.

5.2.3   Methylation and imprinting

Although little is known about the molecular basis of imprinting, the most likely underlying difference is in the degree of methylation of the DNA. The time at which the chromosomes are imprinted is presumably critical and has been suggested to occur during the first meoitic division. It is quite likely that, although overlooked for a long time, imprinting is a fundamental biological process, certainly among the Mammalia.

Bearing in mind that the DNA of the maternal and paternal genomes is indistinguishable, the major question is 'how does the cell know which parent donated which chromosome?'. It must be assumed that there are genes that control this entire process, and these are called imprinting genes, the allele that is affected being the imprinted gene. Inevitably, the greatest interest in this phenomenon results from the medical implications. Details of the mechanism of imprinting have been found in mice. At least three autosomal genes have been shown to be imprinted in mice:

IGF2 Paternal
IGF11 Maternal
H19 Paternal.

The genes, or chromosomes, which are referred to as being paternally imprinted are repressed when inherited from the male. Conversely, maternal imprinting refers to repression of the maternal DNA. Using methylation-sensitive enzymes, such as MspI, to restrict the DNA, it has been possible to show in these genes that there is a difference in methylation depending upon which parent passed on the chromosome.

It has been shown that methylation of CpG islands in close proximity to promoters stops gene expression (see Chapter 3 for more information on CpG islands). In marsupials it is thought that methylation of the X chromosome is not necessary for repression, but the presence of methylation stabilizes the repression. Methylated CpG-binding proteins bind to methylated DNA, thereby denying access to transcription enzymes. This could also account for late replication and inhibited recombination in methylated sequences. Embryologically, the sequential methylation of DNA is of great importance in the control of tissue-specific expression of genes. Mice which are deficient in the enzyme methyltransferase, which is a methylation control enzyme, will not develop to term.

5.2.4 Disrupted development associated with imprinting

The importance of imprinting can be inferred from observations in mice and humans, in which disrupted chromosome distribution can result in two haploid genomes originating from a single parent. Murine fetuses with a completely paternal chromosome set (androgenetic) tend to exhibit poor zygote growth and good placental growth. In the case of a completely maternal genome (gynogenetic) the zygote grows well but the placenta fails to grow properly and early death of the fetus is generally the result. In the human condition the androgenetic zygote forms a hydatidiform mole, a sort of placental tumour. The gynogenetic zygote forms an ovarian teratoma made up of poorly differentiated tissue types.

In terms of viability liveborn human triploids have been recorded. The potential for viability depending on whether there are two paternal chromosome sets and one maternal or two maternal chromosome sets and one paternal. In the case of two paternal chromosome sets and one maternal the placenta is usually highly overgrown but the fetus shows intrauterine growth retardation. Triploids with two sets of maternal chromosomes and one set of paternal chromosomes are not usually viable: miscarriage invariably occurs very early in pregnancy and placental development is found to be very poor.

5.2.5 Clinical aspects of imprinting

Most interest in imprinting has stemmed from disease states, although here again all that can be said is that imprinting exists. In patients with juvenile-onset Huntington's chorea the affected gene is generally paternal in origin. The onset is earlier than when the affected gene is of maternal origin and the deterioration rapid and severe. In a similar way the transmission of myotonic dystrophy via the maternal line will sometimes result in a severe and lethal variant of this disease. Not all suspected cases of imprinting are so easily demonstrated. Paraganglioma, a slow-growing tumour of the head and neck, has been implicated in imprinting of chromosome 11q23–qter.

Prader–Willi syndrome and Angelman's syndrome are of particular interest. The clinical symptoms of these two conditions are markedly different, but both have a prevalence of 1:10 000 to 1:20 000 births. Prader–Willi syndrome is manifested by hypotonia, severe feeding problems, obesity, short stature and hypogonadism. These symptoms may be compounded by later intellectual and behavioural difficulties. Angelman's syndrome is characterized by severe mental retardation, lack of speech, seizures and frequent paroxysms of laughter and stiff jerky movements that account for its alternative name of 'happy puppet' syndrome. For a long time these two conditions were thought to be very closely linked together, but essentially different. An important cytogenetic finding was that nearly all patients studied had a small deletion in chromosome 15q11–13. The difference is, however, that Prader–Willi patients have a deletion in their paternally derived chromosome while patients with Angelman's syndrome have a deletion in the maternally derived chromosome.

Further evidence for the importance of having chromosomes from both parents was provided by the sporadic occurrences of these two syndromes in patients with no detectable chromosome deletions. In these cases it could be shown that uniparental disomy was the cause. This is the condition of having both chromosomes originating from a single parent. If both chromosomes 15 come from the father then the result is Angelman's syndrome. If they both come from the mother then

Figure 5.4 The deletions causing Prader–Willi syndrome and Angelman's syndrome. It should be noted that the syndromes are clinically very different, but the deletion point is the same. The difference in outcome is entirely the result of the parental origin of the deletion. The result of two deletions, one from each parent, is unknown, but is assumed to be inviable.

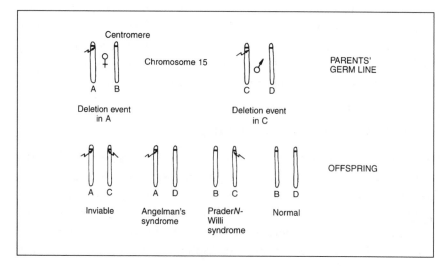

the result is Prader–Willi syndrome. The aetiology of these two syndromes is shown in Figure 5.4. It should be remembered that intuitively it is not possible that two chromosomes transcribing exactly the same product can be differentiated. Therefore it is reasonable to assume that, although Angelman's and Prader–Willi syndrome may appear to be caused by the same chromosome deletion, this cannot in fact be the case. Different, albeit closely linked, genes must be involved. In this particular case it would seem that small nuclear ribonucleoprotein polypeptide N (SNRPN) is associated with Prader–Willi syndrome; however, the gene that is affected in Angelman's syndrome is, as yet, unidentified.

Further evidence for this separation of alleles causing Prader–Willi and Angelman's syndromes comes from families in which there is a normal maternal grandfather, a normal mother and three children all with a submicroscopic deletion. As we know, Angelman's syndrome originates from a maternal deletion; paternal uniparental disomy has the same effect as there is no active maternal allele at the Angelman's site. In the family described above the grandfather has a deletion in the Angelman's region, but on the imprinted chromosome, so that he is of normal phenotype. If the gene was the same as the Prader–Willi gene this would not be the case. Likewise, when this allele is transmitted to the mother it is still an imprinted, deleted, parental allele. But when the allele is transmitted to the next generation the deleted chromosome now shows maternal imprinting. The deleted chromosome is now expressed, the other allele being suppressed by imprinting, so the result is Angelman's syndrome.

Interestingly, not all chromosomes contain imprinted material; uniparental disomy of human chromosome 22 has been shown to have no ill-effects.

5.2.6   Imprinting and tumorigenesis

Imprinting has recently become of considerable importance in the study of various childhood cancers. The evidence of loss of heterozygosity has led to the suggestion that an incorrect imprinting process results in the inactivation of one allele of a tumour-suppressor gene. The converse has also been suggested: that loss of imprinting is associated with gene activation in cancer. One of the mysteries about this process is that loss of heterozygosity or loss of imprinting is nearly always associated with the maternal allele. Although the picture should become clearer with time at the moment it is thought that there may be a tumour-suppressor gene and a growth-promoting gene in close proximity on the same chromosome.

The common finding of spontaneously regressing neoplasms in very early childhood suggests an epigenetic mechanism, such as changes in methylation of the DNA, rather than a mutation event. Mutation events, being well characterized in result, usually do not allow for spontaneous correction of the problem. Retinoblastoma is just such a disease. This condition fits with the two-hit hypothesis of Knudson, which states that both copies of a tumour-suppressor gene need to be inactivated for tumorigenesis to occur. If a tumour-suppressor gene can be imprinted, thereby inactivating it, this would serve as the first hit, the second hit being a deletion or mutation of the other gene. That the events of imprinting can be undone is implied by the spontaneous regression of some cases of retinoblastoma. Other sporadic juvenile tumours fulfilling the predictions of imprinting are Wilms' tumour and osteosarcoma.

The importance of imprinting in adult human cancers may become apparent as studies progress. Chromosome loss in human tumours such as osteosarcoma and Wilms' tumour tends to be maternal, whereas the chromosome 9 involved in production of the Philadelphia chromosome, which is so frequently associated with leukaemias, is nearly always paternal in origin.

Although attempts have been made to link an imprinting effect with an abnormal phenotype in cases of balanced translocations, this has not proved to be a fruitful line of investigation.

5.2.7   X chromosomes in mammals

One of the commonest forms of imprinting is the inactivation of the X chromosome in mammals. In marsupials it is the paternal X chromosome which is preferentially inactivated. The very specialized mechanism of X-chromosome inactivation will be dealt with in more detail in Chapter 8, since this has considerable implications for sex determination.

Just as DNA methylation is important in autosomal imprinting, transcriptional control of whole chromosomes can also be seen as imprinting. In *Drosophila*, X-chromosome equalization is achieved by raising the output of the hemizygous male X. This is achieved by acetylation of histone H4 lysine. Conversely, in mammals in which compensation is required this is achieved by silencing one of the X chromosomes. The quiet X chromosome has very low levels of histone H4 acetylation while at the same time the DNA is hypermethylated.

Candidates for imprinting genes have been sought although none has as yet been positively identified. It has been suggested that imprinting genes may be either histone or non-histone protein genes. It is unlikely that histone genes are involved in any but a peripheral way, but non-histone protein genes may well be. The difference between euchromatin and heterochromatin is in the protein component rather than in the DNA sequence. The process of imprinting, or gene inactivation, will no doubt require more than one gene product. The areas of the X chromosome which are not susceptible to inactivation may well turn out to harbour at least one imprinting gene.

The question must arise as to why imprinting occurs at all since when it goes wrong it has such disastrous results. Although there are several hypotheses as to the reason for imprinting, it is worth making the point that none of them has been proved and it is conceivable that the reason may be different in different species.

Reasons for imprinting fall into three categories. The first suggests that it is associated with sex determination. If there were two sexes but no sex chromosomes, imprinting would perhaps allow differentiation between the sexes; thus, evolution of sex could have occurred before evolution of sex chromosomes. The second major reason revolves around the notion that dominance and inappropriate cross-fertilization can be managed using imprinting to create non-viable zygotes. However, the most likely hypothesis to account for imprinting is that it is important in embryogenesis. Evidence for this comes from the known consequences to embryos which have received the entire genome of one or other parent, as previously described. In mammals the fetus would be competing in the uterine environment for resources, so excessive development of either the embryo or placenta could be prevented by the maternal or paternal genomes.

One aspect of imprinting which must be considered is how the cell recognizes the parental origin of genes and chromosomes. This can only be achieved by having *cis*-acting sequences which mark the alleles. It is certain, however, that the detailed mechanism of imprinting is a complicated one involving several methods, both local and regional, to control individual genes, clusters of genes and whole chromosomes.

5.2.8   Evolutionary origin and causes of imprinting

Many reasons for imprinting have been suggested. One such suggestion is that it forces sexual reproduction on eukaryotic species. This does not take into consideration the fact that asexual reproduction can be a positive advantage and that some species are completely parthenogenic. In a similar vein, the suggestion has been made that imprinting maintains diploidy by incapacitating monosomic cells.

At this time the most likely explanation for the origin of imprinting would seem to be dosage control. Most genes are not imprinted, as can be inferred from the straightforward mendelian inheritance of most genetic characters and genetic diseases. From this it can be assumed that either dosage is not important or that both copies are required of most genes. In the case of imprinted genes it is important that only a single gene is active; by switching off one gene a local state of hemizygosity is induced. We can clearly see that the result of a double dose of some genes is deleterious, as in the case of embryogenesis referred to above. But it must be remembered that it is a dosage effect alone; there is no such thing as male and female DNA. Imprinting occurs anew at every generation.

Mutagenesis

6

Summary

We explore here the issues of mutagenesis, specifically where they apply to chromosomes. Points to note are that mutagenesis is a process that goes on all the time. Errors of replication result in a continuous burden of mutations that have to be dealt with by cells. Another interesting natural form of mutation is the transposable element, which although usually dormant can produce interesting phenotypic effects when expressed.

Chemical mutagens and radiations increase the genetic load, but it is not generally possible to identify the cause of any mutation from the result alone. For an organism, the effect of small-scale mutations can be devastating, as in oncogenesis.

For a species, uncontrolled chromosomal rearrangements resulting in non-viable recombination products are disastrous. Reproductive viability, in this case, would be so reduced as to endanger survival.

6.1 General ideas about mutagenesis

Although mutagenesis can sometimes appear to occur spontaneously, it should be remembered that biological systems obey the laws of thermodynamics in the same way as any chemical system. Therefore there is always a reason, a causal agent, for errors of replication and changes within the genome, although it may never be possible to be sure of the cause in any specific case. When considering large numbers of

Table 6.1 Spontaneous mutation rates for organisms in a population. These data do not take account of variations within individuals or between individuals

Species	Mutation rate	
Escherichia coli	2×10^{-8}	per cell division
Zea mays	1×10^{-6}	per meiotic division
Chlamydomonas spp.	1×10^{-6}	per cell division
Neurospora crassa	6×10^{-8}	per asexual spore
Drosophila spp.	4×10^{-5}	per meiotic division
Homo sapiens	3×10^{-5}	per meiotic division

individuals, or populations, it becomes easier to quantify this apparently spontaneous mutation rate which is due to environmental factors met with on a day-to-day basis. Such calculated mutation rates for organisms do not take into account the type of mutation, whether it is a point mutation or a much larger event, or the fact that genes differ in their robustness when confronted with environmental insults.

Table 6.1 shows some of these overall mutation rates. The three broad groups of error which can occur within a gene to alter a function

Figure 6.1 Three main types of error can occur to a gene to alter function. (a) A single base change alters a single amino acid in the final product. This is shown in the case of the *RAS* oncogene. (b) Deletion or insertion of one or two base pairs causes a frame shift. This makes a profound difference to the product, which can result in abnormal chain termination. (c) Deletion or insertion of multiples of triplets results in an abnormal product with amino acids being either added or missing.

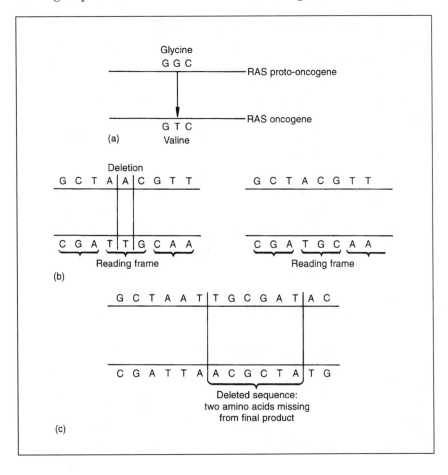

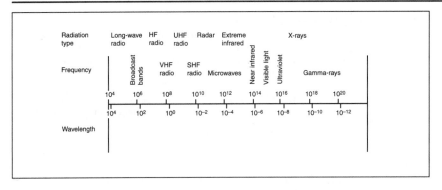

Figure 6.2 The electromagnetic spectrum. It is interesting to note that, although we operate using visible light, such radiation constitutes a tiny range of the complete spectrum.

are shown in Figure 6.1. Changes in gene switching resulting from massive translocations are dealt with later.

Once they have occurred it is not possible to tell from the final outcome whether a mutation event is the result of a chemical mutagen, such as benzene, or ionizing radiation, such as X-rays, or non-ionizing radiation, such as ultraviolet radiation. This is not the case with transgenic mutagenesis however, in which case the host gene is implicitly marked. Small-scale mutation events, usually involving single base changes, such as thymine dimer production by UV light or the conversion of benzpyrene into 5, 6-epoxide by the liver enzyme aryl hydroxylase, are not really within the scope of this discussion, so we will confine ourselves to larger mutations. These are ones which involve genetic material at the chromosome level.

Figure 6.2 shows the range over which radiations are found and their energies. Table 6.2 is an an exact equivalent of this but for chemical mutagens, grouped generically.

Radiations can alter gene expression without damaging the gene

Base analogues	These are chemicals which are sufficiently similar to the normal purines and pyrimidines that they can be incorporated into synthesizing DNA. This can cause a base pair mismatch. Examples are Bromodeoxyuridine and Bromouracil
Alkylating agents	These react with the bases, adding ethyl or methyl groups. This can result in a gap or a pairing mismatch. The base most often affected is guanine. Examples are nitrogen mustards and ethyl methane sulphonate
Intercalating dyes	These chemicals lodge between adjacent bases and distort the helix. This causes deletions or additions during replication to maintain the overall integrity of the helix. Examples are acridine orange, quinacrine hydrochloride and bisbenzimidazole dyes
Deaminaters	These chemicals act, as their name indicates, by deaminating the bases. This results in mismatching of base pairs. An example is nitrous acid

Table 6.2 The four major classifications of chemical mutagens.

Box 6.1 Absorbed dose of radiation

The absorbed dose is defined as the energy deposited in a medium by a radiation. Originally this was measured in rads and was the energy deposition of 0.01 J kg^{-1}. The rad has been replaced by the gray (Gy). This is defined as 1 J kg^{-1}. Thus:

$$1 \text{ Gy} = 1 \text{ J kg}^{-1} = 100 \text{ rad}$$

Biological systems are affected differently by the same physical dose of different radiations. This is shown by the equivalent amount of damage done to living cells by two different radiations:

$$1 \text{ Gy (100 rad) gamma-radiation} = 0.05 \text{ Gy (5 rad) fast neutrons}$$

A quality factor (*Q*) is then introduced:

$$\text{Absorbed (Gy)} \times \text{quality factor } (Q) = \text{dose equivalent}$$

The value assigned to *Q* varies such that:

Radiation type	Value of *Q*
X-rays	1
Gamma-rays	1
Electrons	1
Thermal neutrons	5
Fast neutrons	20
Protons	20
Alpha-particles	20

itself; obvious examples of this are the production of melanin in human skin on exposure to ultraviolet radiation and production of chlorophyll by plants in response to light. Another, animal, example is the control of a stress response protein, heme oxygenase 1, which is produced in large amounts when cells are exposed to ultraviolet radiation. This is caused by a change in transcription rate due to the irradiation of controlling factors, rather than a direct result of irradiation of the gene itself (Box 6.1).

Before looking in more detail at the mechanisms and results of mutagenesis it is worth noting that the disease burden resulting from inherited defects, both those due to single-gene defects and those of multifactorial origin, is quite considerable at 10%. A breakdown of the causes of inherited diseases is given in Table 6.3.

6.2 Changes associated with the environment

The background rate of chromosome damage in human somatic cells, caused by environmental wear and tear, is rather interesting in that,

Dominant and X linked	1.0
Recessive	0.1
Malformations of unknown aetiology	4.3
Multifactorial	4.7
Chromosome abnormalities	0.4
Total	10.5

Table 6.3 Percentage of human births affected by congenital malformations at birth. Data from United Nations Scientific Committee on the Effects of Atomic

although it would seem reasonable to assume breaks and rearrangements to be the commonest form of aberration, this is not in fact so. Breaks and rearrangements are present in up to 1% of lymphocytes, whereas the rate of aneuploidy in lymphocytes, although it varies with age, sex and other factors, e.g. smoking, can be up to 12% without any detrimental effect. It should be remembered that these data were obtained in cultured lymphocytes and that the process of *in vitro* culture can induce artefactual aneuploidy.

Mutations which are the result of foreign DNA being inserted into the genome are rather different in that they do not usually affect the chromosomes directly. These mutant forms are normally induced by introducing DNA into either embryonic stem cells or directly into embryos. Such methods have been used extensively in mice to cause them to express foreign proteins.

It has been pointed out in the past that there is an anomaly in the response of different genes and different points within genes to similar environmental stress. This anomaly is that, although the long-term structure of DNA is ostensibly random, the susceptibility to mutation events is most definitely not.

The two commonest types of mutation affecting genes, point mutations and length mutations, each account for approximately half of known cases. An explanation of non-random events is that methylated gene sequences seem to be particularly prone to mutation. Methylcytosine is able to undergo deamination to thymine with the corresponding strand base (G) being replaced with an A. Length mutations also seem to be localized within specific areas: most deletions seem to occur where a repeat sequence is associated with the gene. In animals enzyme systems can create carcinogenic compounds from relatively common environmental products (Box 6.2).

It has been suggested that the only way that a large, complex and long-lived organism can survive is to have an immune system which can recognize mutations within the body. This would then be the primary activity of the immune system, with defence against invading organisms being a secondary function.

Box 6.2   Innocent chemicals and dangerous metabolites

Apparently safe chemicals can sometimes be metabolized to new and highly toxic or mutagenic products. A method that is increasingly being used to circumvent this problem is to alter the original chemical slightly such that its primary use is unaffected, while its metabolic products are far less toxic than those of the original chemical.

An example of such a chemical is ethylene glycol, whose breakdown products are toxic to humans.

Metabolism

$$R\text{-O-CH}_2\,CH_2\,OH \longrightarrow R\text{-O-CH}_2CH = O \longrightarrow R\text{-O-CH}_2C = O$$
$$\underset{\text{OH}}{\big|}$$

Ethylene glycol ether   Alkoxy aldehyde   Alkoxy acetic acid

Modifying the starting material slightly alters this pathway without affecting the useful properties of the original chemical and results in less toxic metabolites.

Metabolism

$$R\text{-O-CH}_2CHOH \longrightarrow R\text{-OH} + HO\text{-CH}_2\text{-CHOH}$$
$$\underset{\text{CH}_3}{\big|} \qquad\qquad\qquad\qquad\qquad \underset{\text{CH}_3}{\big|}$$

1 (Alkoxy) propylene   Alcohol + propylene glycol

There are many examples of procarcinogens, chemicals which are themselves relatively harmless but are transformed into powerful carcinogens by enzymes found in the body. This at least in part helps to explain why some species appear to be immune to toxins or carcinogens: they simply do not have the cellular machinery required to convert the harmless into the dangerous.

Simple examples of this process are shown below.

Procarcinogen	Liver enzyme	Carcinogen
Benzpyrene	Aryl hydroxylase $\longrightarrow$	5,6-Epoxide
Aflatoxin	Aryl hydroxylase $\longrightarrow$	2,3-Epoxide

A more complicated, multistep, example is the *in vivo* conversion of nitrates, themselves of no mutagenic potential, into very powerful carcinogens called nitrosamines. This is shown in Figure 6.3.

Figure 6.3

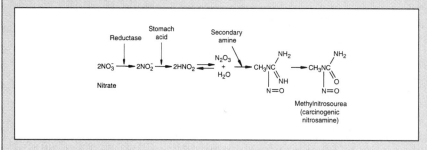

6.3   Chromosomal aberrations

Chromosomal aberrations fall into three groups: aneuploidies, deletions and rearrangements. Such chromosomal changes occur naturally,

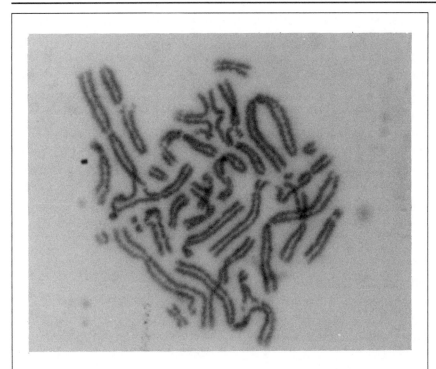

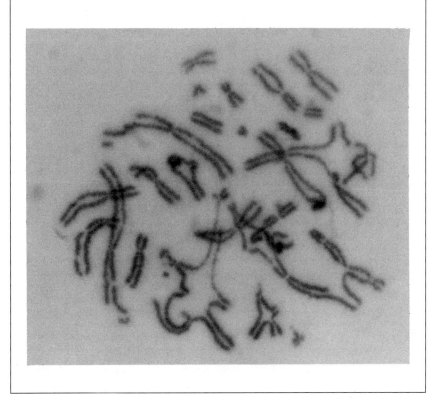

Figure 6.4 Restriction enzyme damage to human cells. Photograph courtesy of A.H. Harvey.

sometimes as a part of evolution, so when studying mutagenic changes it should be remembered that these represent not a new phenomenon but an increase in the frequency of a natural process. As a cell goes through cell division chromosome damage is usually the first visible sign of a problem caused by mutagenesis. Indeed, chromosome damage is one of the ways in which bacteria can protect themselves from foreign DNA. Bacteria produce restriction endonucleases, which cut specific methylated sequences by acting on an internal phosphodiester bond. Section 3.2.6 describes how the reaction of endonucleases with fixed metaphases can be put to good use. If, by contrast, the enzyme is introduced into growing cells, the chromosome damage can be considerable, causing cell death by complete disruption of cell division. Just such damage is seen in Figure 6.4.

It has been shown in higher plants that there is a relationship between the radiation dosage required to kill a plant and the amount of DNA present in each chromosome, the most sensitive species having a few large chromosomes and the most resistant many small ones. Nowadays radiation is used to beneficial effect in the radiation of foodstuffs (Box 6.3).

It can be said with some certainty that 0.4% of liveborn humans have a chromosomal abnormality of some form. It has been suggested that against this background the additional genetic abnormalities in the form of aneuploidies and translocations in newborns which are attributable to parental radiation exposure are most likely quite unimportant, although damage to the individual who receives the irradiation might be quite severe. Of course, if very specific areas are irradiated then high doses can be tolerated without killing the individual. By irradiating the spermatogonia of mice it is possible to induce large deletions of 2.5–30% of progeny chromosomes while still maintaining viability and fertility.

When assessing radiation damage it seems obvious that, for any given species with a constant number of chromosomes, the mean number of hit chromosomes will be constant at any given dosage. However, if the dosage is doubled, thereby doubling the number of hits, the number of hit chromosomes will not double as some chromosomes will receive multiple hits. For example, in a species with six chromosomes, a cell can only receive a maximum of six hits before some chromosomes will have been hit twice. The issue of single or multiple hits to a nucleus is important because different aberrations require different numbers of hits to induce them. Thus, we can demonstrate that the relationships follow simple geometric rules relating dosage (hits) to effect (aberrations). For an aberration induced by a single hit:

Box 6.3 Irradiation of foodstuffs

The use of irradiation of foodstuffs as a means of preservation has two main advantages: it is cold and non-chemical. Irradiated foodstuffs are identified by specific labelling. There are three main types of radiation which can be used for this process: gamma-rays, X-rays and electrons.

Electrons have very low penetrating power and therefore can only be used to irradiate thin films, such as polymer coatings, or liquids. Although irradiation is regarded as cold, very small increases in temperature do occur, as would be expected from the absorption of energy.

If electrons are stopped by a lead target then X-rays are produced, although with a considerable loss of energy in the conversion. These can then be used for X-ray irradiation.

To avoid inducing unacceptable levels of radiation in the product energies are kept below certain values.

10 MeV for electrons
5 MeV for X-rays
2 MeV for gamma-rays.

Induced radiation decays about 10- to 20-fold in the first 24 h after treatment although it is alleged that induced radiation never achieves levels higher than the natural environmental radiation due to carbon-14 and potassium-40.

Although food irradiation is often treated with suspicion, it should be remembered that large quantities of biomedical equipment are sterilized using radiation. Other than ethylene oxide gas, irradiation is one of the surest ways to guarantee sterility of equipment that cannot be autoclaved.

The most common source of radiation is cobalt-60, which is produced by irradiating cobalt-59. Occasionally caesium-137 is used, originating from spent nuclear fuel.

The first patents for food irradiation were granted in the early twentieth century for the use of X-rays.

The first commercial use of irradiation as a preservative was to stop potatoes from sprouting. This was allowed in 1958 in what was the former USSR and then in 1960 in Canada. In 1959 the USSR started irradiating grain to eradicate insect pests.

Sprouting is prevented by irradition because the DNA is damaged, resulting in cessation of cell division but not the normal physiological processes. Radiation can create radicals which damage large molecules, such as DNA, and sensitive molecules, such as vitamins. Vitamins which can be depleted by irradiation are B_1, C and E.

Some doses which are used for specific purposes are shown below. The lowest doses are used for sprout inhibition and the highest for sterilization.

Approximate dose (kGy)	Result
0.1	Inhibition of sprout growth
0.15–0.5	Insect deinfestation
2.0–5.0	Micro-organism erudication
30–50	Sterilization

It is worth remembering that one thing cannot be more sterile than another; it is either sterile or it is not!

$$\text{Aberrations} = kD + C$$

where k is a constant, D is the dosage and C is the basal aberration rate in non-irradiated cells. It can be seen that this equation is of the same form as the simple equation for a straight line:

$$y=mx+c$$

Thus, if two hits are required to cause a specific aberration then the number of aberrations will be proportional to the square of the dose. This can be expressed simply as:

$$\text{Aberrations} = kD^2 + C$$

As the number of hits required increases so the formula can be altered accordingly; so for a three-hit aberration the formula becomes:

$$\text{Aberrations} = kD^3 + C$$

It should be noted that C is usually so small that it is ignored. Table 6.4 lists the relationships for several specific aberrations in the plant *Vicia faba*.

A general point to make here is that whenever there is a rearrangement of chromosomal material it is likely to affect the organism concerned in some very practical way. In somatic cells the resulting effect may be small, even when the changes are relatively large, reflecting the fact that most somatic cells are fully differentiated and non-dividing. When changes occur in a gamete, however, the resulting effects may be devastating, rendering the zygote inviable.

As stated earlier, the distribution of gene mutations is not random. In contrast, there is a growing body of evidence that chromosome aberrations are randomly distributed. In the case of mutations affecting cell proliferation it is possible to conclude that the events which cause the problem are not random. In some cases, however, the distribution of chromosome breaks and rearrangements can be clearly seen to be random. In a study of the distribution of chromosomal aberrations in natural populations of *Drosophila nasuta* 85 inversions were found: eight on chromosome X, 24 on chromosome 2 and 53 on chromosome 3. Although this does appear to be non-random it should be remembered that many factors will determine whether any mutation or change becomes fixed in a population. Damage induced by radiation in this species was randomly distributed among all chromosomes, which would certainly imply that, whatever the final outcome of a mutation event, the chromosomes are equally sensitive to radiation damage.

Table 6.4 Relationship between X-ray dosage and type of chromosomal aberration induced in root meristems of *Vicia faba* (Revel 1966)

Gaps	Yield = $kD^{1.07}$
Chromatid interchange	Yield = $kD^{2.0}$
Isochromatid deletion	Yield = $kD^{1.46}$
Chromatid deletion	Yield = $kD^{1.71}$
Triradials	Yield = $kD^{2.92}$

6.4 Aneuploidies

Aneuploidies can be induced in virtually any cell or organism, although not all chromosomes will be involved with the same frequency. Generally, the larger chromosomes are involved less often than smaller ones, although in the case of humans the true picture is masked. Human chromosome 19 is the most frequently lost chromosome in cultured fibroblasts, but this is not reflected by the prevalence in livebirths because, as in the case of most human aneuploidies, it results in non-viability. In humans trisomic fetuses are viable in the case of only five chromosomes: 13, 18, 21, X and Y. Most known cases of aneuploidy occur in humans, although the precise method by which aneuploidies originate is unknown. In the case of polysomies which involve multiple sets of sex chromosomes, the additional sets are always uniparental in origin; for example, in the case of 46, XXYY the additional XY will have originated from one parent. One of the best-known human autosomal aneuploid conditions is Down's syndrome. This is due to the presence of a complete extra chromosome 21. Although in humans, Down's syndrome results in severe mental retardation and other problems there is a described case of trisomy 21 in a gorilla in a German zoo which was found by chance. This particular animal was 5 years old and showed no abnormalities, suggesting that the overall activity of a species' genome is dependent on more than merely a sequence.

There is no doubt that the risk of Down's syndrome increases with

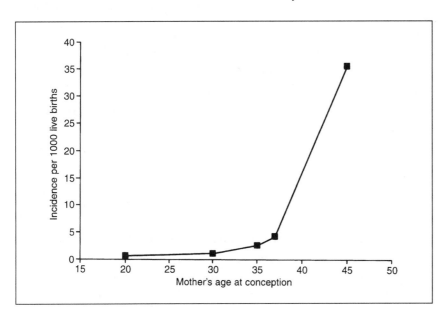

Figure 6.5 Maternal age effect on the incidence of Down's syndrome in humans.

Figure 6.6 C-banded metaphase of human chromosomes showing a fusion between chromosomes 14 and 21. Although there are only 46 chromosomes present in the cells of this individual, there are three copies of chromosome 21. Courtesy of S.C. Rooney.

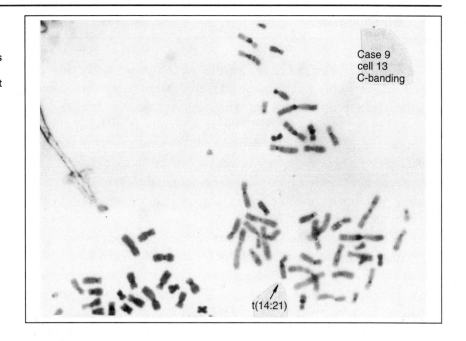

Case 9
cell 13
C-banding

t(14:21)

maternal age, as shown in Figure 6.5. This is not the entire story, however, because there are in fact two types of Down's syndrome: one in which there is an additional free chromosome and one in which a translocation product results in the transmission of the extra chromosome 21 fused to another chromosome. This latter group mostly comprises Robertsonian fusions in which the telocentric chromosome 21 fuses to another telocentric chromosome, either 13, 14, 15, 21 or 22. Just such a translocation is shown in Figure 6.6. In this particular case the fusion is between chromosomes 14 and 21.

These fusions in themselves can have interesting outcomes. For example, if the chromosome 21 fuses with a chromosome 13 then it is possible for a zygote to have an additional chromosome 13 (Patau's syndrome) or an extra 21 (Down's syndrome). If the 21 fuses with, say, a chromosome 15, which is also telocentric, then the zygote that has the additional chromosome 15 will not survive; not all chromosomes of similar sizes are of the same importance for the organism's survival.

The frequency of different types of chromosomal anomalies in livebirths is shown in Table 6.5.

In the case of Down's syndrome it has recently become possible by using DNA markers to show that in some 4.5% of cases of Down's syndrome in which the extra chromosome is unattached to another chromosome, the error of nuclear division took place not during

Anomaly	Rate (%)
Polyploidy	0.013
Autosomal trisomy	0.124
Monosomy X	0.0046
Other	0.443
Total anomalies	0.623

Table 6.5 Rate and type of chromosomal anomalies found in livebirths

meiosis, which is the case with Robertsonian fusions, but during mitosis. Such errors may be post-zygotic, occurring in very early embryogenesis, or during the development of the parental gonads.

Turner's syndrome, which is the result of monosomy of the X chromosome, occurs in 1 in 5000 live female births. This figure is considerably less than the estimated 1–2% of all conceptuses which are thought to be affected in this way. Exactly why so many such affected fetuses fail to reach term is unknown; there does not seem to be a single reason. The mechanism resulting in monosomy X is as yet unexplained, but meiotic non-disjunction is probably not the cause, as there is no correlation with maternal age. The Turner's syndrome phenotype varies quite widely and is at least in part a reflection of the level of involvement of the X chromosome. Complete monosomy results in the classical syndrome, but partial monosomies and even cases in which the karyotype is ostensibly XY are known. In this last case there is an inevitable loss of genetic material from the very tip of the short arm of the Y chromosome. Ring chromosomes, which are formed when telomeric material is deleted and the resulting ends of the chromosomes become 'sticky' and bind to each other, will result in Turner's syndrome variants. In this case the severity of the condition reflects the amount of DNA which has been lost. Whereas the Y chromosome does not seem to represent very much in the way of active DNA and therefore only the p arm is significant when considering sex determination, in the case of the X chromosome any loss seems to cause some phenotypic disturbance. This is the only complete monosomy that is compatible with post-natal life, presumably reflecting the strange inactivation mechanism involved in mammalian sex determination (Chapter 8).

There are few animal models of Turner's syndrome. The animal usually studied is the mouse, which remains apparently normal and is fertile. There are, however, examples of horses, pigs and cats with XO karyotypes, all with severe skeletal defects and shortened lifespan.

6.5   Mutations affecting cell proliferation

Under normal conditions cells are only able to proliferate for a fixed number of cell divisions. The exact reason for this is unknown, but generally the number of divisions a cell can undergo in culture is very similar to the number of divisions a cell can undergo in the original host. Thus, human fibroblasts tend to proliferate for about 50 cell divisions before the culture dies out. Transformed cell lines are different in that, once immortalized, the cells can continue to divide indefinitely. Certain properties of transformed cells are relatively easy to define; they lose their ability to adhere to their neighbours and cell division is not inhibited by proximity to other cells.

A change as profound as the immortalization of a cell line, either *in vitro* or *in vivo*, requires a fundamental change in the genetic make-up of the cell. Given that a mutation event has taken place, the questions which remain are 'what is the change which takes place within the cell?' and 'how does it manifest itself at both the chromosomal and molecular level?'. The first question has been investigated and certain consistencies have become apparent.

Leukaemias and lymphomas are rather better characterized in terms of chromosomal rearrangements than solid tumours, as can be seen in Tables 6.6 and 6.7. Surprisingly, some of the changes which have been detailed are remarkable in being both consistent and relatively simple.

One of the problems of determining the relevance of a chromosomal change with respect to mutagenesis is whether the change is causal or simply a degenerative result of immortalization of a cell line.

Many chromosomal changes that take place in cultured cells reflect the lack of need for accurate control of genetic elements once a cell is

Table 6.6   The precise breakpoints and genetic lesions in well-characterized leukaemias

Translocation	Leukaemia type	Genetic lesion
t(15;17) (q21;q11–12)	Acute myelogenous leukaemia	N-terminus of PML (zinc finger protein) attached to C-terminus of retinoic acid receptor α-gene
t(1;19) (q23; p13)	Acute pre-B cell lymphoblastic leukaemia	PBX (homeobox gene) on chromosome 1 attaches to E2A, an immunoglobin enhancer-binding protein
t(17;19) (q22;p13)	B-cell acute lymphocytic leukaemia	Hepatic leukaemia factor (HLF) translocated onto E2A gene

	Chromosome																							
Tumour type	1	2	3	4	5	6	7	8	9	10	11	12	13	14	15	16	17	18	19	20	21	22	X	Y
Bladder	X		X		X		X		X		X													
Breast	X						X									X								
Colon	X				X	X											X	X						
Kidney					X									X										
Ovary	X					X	X				X			X										
Prostate							X			X														
Testicular												X												
Lung			X																					
Synovial																		X					X	
Melanoma	X					X			X										X					
Nuroblastoma	X																							

Table 6.7 Some of the chromosomes consistently seen to be altered in solid tumours. The involvement may be in the form of translocations or deletions

proliferating out of control. Accumulated errors become more apparent and more tolerated, but it is not immediately obvious whether these changes are the cause or a reflection of mutagenesis.

6.6 *Translocations*

It has been known for some time that oncogenes are a primary cause of malignancy. These genes, once thought to be only elements of destruction, are now known to be fundamental to the development of an organism. They are highly conserved across phylogenetic groups and are normally switched on and off in a very precise way. When they are switched on in the wrong place they cause fundamental problems to the organism. The products of these abnormal switches can be very precisely defined and are often associated with specific age groups and tissues. Table 6.8 shows some of these very specific alterations.

It should be remembered that translocations are not always detrimental: there is a long tradition in classical cytogenetics of benign translocations being found. Almost all possible rearrangements have been found at one time or another. Whether these translocations are artefacts of tissue culture, or perhaps of the processes necessary to produce metaphase spreads for microscopy, or true reflections of the

Table 6.8 Chromosomal aberrations in soft-tissue tumours in humans

Carcinoma type	Affected tissue	Chromosomal association
Liposarcoma	One of the commonest soft tissue cancers in adults	t(12;16) t(12;16) + 8
Synovial sarcoma	A sarcoma of adolescents, mostly of the knees	t(X;18)
Rhabdomyosarcoma	This is the commonest sarcoma in under 25s. Mostly of the head, neck and genitourinary system	t(1;13) t(2;13)
Ewing's sarcoma	The second commonest bone malignancy in adolescents. Mostly of the long bones, but sometimes of the soft tissues	t(11;22) t(1;16)
Chondrosarcoma	Sarcoma of cartilage	t(9;22)
Mesothelioma	Sarcoma of the pleura or peritoneum mostly in 50 to 70-year-olds	del(3) del(9) del(6)
Desmoplastic small round cell tumour	This tumour is mainly intro-abdominal	t(11;22)
Fibrosarcoma	This affects the distal regions of extremities in infants and children	+ 11, + 20 + 11, + 20, + 17

Soft-tissue tumours are a heterogenous group of disorders. Diagnosis is problematic in these disorders even using all forms of histological techniques. Cytogenetics is useful here as a consistent picture is starting to emerge of chromosomal changes associated with specific sarcomas. It should be remembered that these consistent changes are frequently masked by other, degenerative changes.

individual organism is not always clear, but there are very few which are intrinsically harmful. Translocations which are known to be detrimental to the carrier are the well-documented translocations involving oncogenes and cases of a gene being incorrectly activated or inactivated as a result of a change in its location within the genome.

In the special case of defective repair mechanisms, breaks and rearrangements can eventually lead to the production of a clone of malignant cells. Some examples of such conditions are described.

1. In Fanconi's anaemia the chromosomes break, contain gaps and produce multicentric chromosomes and acentric fragments. Acentric fragments are lost from the cell, as would be expected as they have no centromeres. This is a rare autosomal recessive disorder which is manifested between the ages of 4 and 12 years. Although

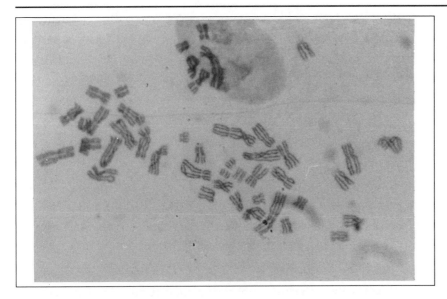

Figure 6.7 Endoreduplicated human chromosomes from cultures fibroblasts. The chromosomes replicate but are unable to divide correctly. Courtesy of A.N. Harvey, MRC Radiobiology Unit.

it causes various skeletal and pigmentation anomalies it is the chromosomal anomalies that are of particular interest. Of these, endoreduplication is the strangest phenomenon, occurring in both lymphocytes and fibroblasts. Figure 6.7 shows the result of endo-reduplication in cultured human fibroblasts at metaphase. The chromosomes have replicated but are unable to separate correctly at anaphase, thus resulting in this abnormal configuration.

2. The chromosomes of patients with Bloom's syndrome exhibit all the same rearrangements as in Fanconi's anaemia but are also

Figure 6.8 Sister chromatid exchange (SCE). (a) A normal human fibroblast cell line with the chromatids distinguishable by staining. Very few exchanges can be seen. (b) Metaphase from a Bloom's syndrome patient. The very high rate of SCE can be recognized from the harlequin pattern. Courtesy of A.N. Harvey, MRC Radiobiology Unit.

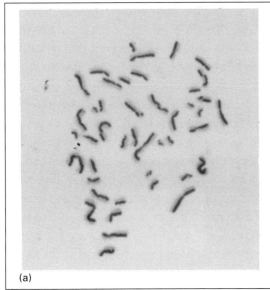

(a)

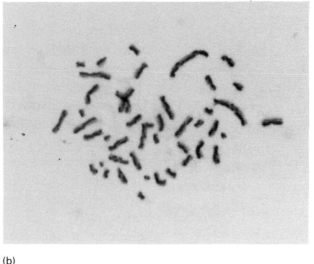

(b)

characterized by the production of symmetrical quadrivalent chromosomes. One feature of Bloom's syndrome that is of particular interest is the very high rate of spontaneous sister chromatid exchange. Figure 6.8 shows this admirably. The chromosomes are grown in BrdU so that the chromatids can be distinguished. When sister chromatids are exchanged the result is a harlequin pattern.

3. In ataxia telangiectasia, the same errors as in Fanconi's anaemia occur but with a much reduced frequency. Interestingly, lymphocytes from these patients tend to respond poorly to mitogenic stimulants such as phytohaemagglutinin. This is an autosomal recessive condition. The most interesting chromosomal changes which occur on a regular basis are translocations involving chromosomes 7 and 14. Such rearrangements are seen as a normal consequence of tissue culture procedures but at a very low frequency.

It has been shown that these chromosomal repair problems are not artefacts of culture but a genuine reflection of what is occurring in the patient. It can be assumed from this that it is the breakage and inadequate repair of the chromosomes which result in the symptoms of the disease. Various mechanisms of environmental insult can increase the basal rate of chromosome damage in these patients, both chemical mutagens and ionizing radiations (Box 6.4).

In the case of the autosomal recessive disorder, ataxia telangiectasia, while the homozygous individual is severely affected, the heterozygous carrier also seems to have an increased risk of cancer. When intrachromosome recombination rates were compared in ataxia telangiectasia cell lines and control xeroderma pigmentosa cells and normal human fibroblasts, it was found that the ataxia cells exhibited elevated rates of recombination. This could explain the elevated rate of cancer in patients with this disease perhaps, as a result of loss of heterozygosity at an oncogene locus.

There is a general point to be made here about crossing over rates in chromosomes involved in translocations. Both intrachromosomal crossing over (involving the same chromosomes) and interchromosomal crossing over (involving different chromosomes) seem to be affected by the presence of translocations within a nucleus. This phenomenon does, however, seem to be species-specific. In many species there is evidence of increased recombination of translocated chromosomes. Interchromosomal effects are well known in *Drosophila* but in Robertsonian fusions of chromosome 21 in humans no effect has been found.

Box 6.4 Chernobyl

The Chernobyl disaster occurred on 26 April 1986, at 01.23 local time.

In 4 secs power increased one hundred times above normal in Chernobyl reactor number 4. This caused an explosion which released as much radioactive material as 90 Hiroshima bombs over a period of 10 days.

After the accident reactor number 4 was encased in a concrete tomb 60 m high, 60 m long and between 6 and 18 m thick. As a result of continuous radiation exposure and a strong thermal gradient from inside to outside, the concrete started breaking up within 5 years.

In the surrounding area there are approximately 800 pits lined with 10 cm of clay containing all the waste, both domestic and industrial, collected during the clean-up process.

The three main components of the fall-out were:

^{131}I, which accumulates in thyroid glands;
^{90}Sr, which accumulates in bones;
^{137}Cs, which accumulates throughout the body.

These isotopes are all electron emitters.

^{131}I was regarded as a temporary problem since its half-life is only 8.04 days. In contrast ^{90}Sr has a half-life of 28 years and ^{137}Cs has a half-life of 30 years. The table below lists some isotope half-lives which serve to indicate the length of time for which reactors may need to be supervised.

Isotope	Half-life
^{232}Th	1.39×10^{10} years
^{238}U	4.5×10^{9} years
^{40}K	1.4×10^{9} years
^{235}U	7.1×10^{8} years
^{226}Ra	1620 years
^{14}C	5700 years
^{60}Co	5.25 years
^{237}Am	1.3 hours

It should be remembered that isotopes with a short half-life are disintegrating so fast that very small amounts will produce high levels of radioactivity. For example, shown below are the masses of pure isotope in a source of 1 μCi.

Isotope	Mass of 1 μCi (g)
^{238}U	3.0
^{226}Ra	1.0×10^{-6}
^{90}Sr	7.0×10^{-9}
^{60}Co	8.8×10^{-10}
^{131}I	8.1×10^{-12}

In UK the maximum acceptable dose of radiation is 0.5 mSv per year. In Chernobyl the acceptable lifetime dose is 350 mSv. This is equivalent to a ground dosage of approximately 15 Ci km^{-2}.

Areas west of the plume received fall-out of up to 2667 Ci km^{-2}. Areas with more than 40 Ci km^{-2} K were evacuated in 1986 and fenced off.

The town of Poliske was found to have maximum levels of 370 Ci km^{-2} with an average of 26 Ci km^{-2}. Of 5670 children examined, 2440 were found to have enlarged thyroids as a result of being exposed to ^{131}I radiation in excess of 2 Sv (200 rem). In nine children doses of ^{131}I exceeded 10 Sv.

The death rate among the decontamination workers has been very

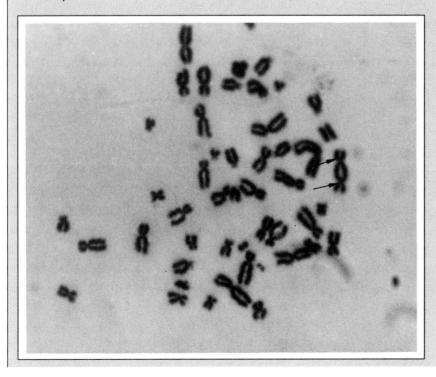

Figure 6.9 Dicentric chromosome induced by X-rays in a cultured human fibroblast. Courtesy of A.N. Harvey, MRC Radiobiology Unit.

high, especially among those who were involved in cleaning the debris from the contaminated roof of reactor number 3.

Overall the region of Belarus received 70% of the fall-out and has lost 20% of the cultivatable land because of contamination.

As a result of retention of caesium by lichens eaten by reindeer, 80% of reindeer slaughtered in 1986 were unfit for human consumption. Long-term retention and mobility within the soil cause long-term problems. For the same reason, 5 years after the accident the sale of sheep from some upland farms in the UK was still banned.

The long-term effects of Chernobyl may not become apparent for many years, as the incidence of cancers changes with long-term dosages.

Studies were carried out on the chromosome aberration rate 5 years after the Chernobyl accident was studied in the population of Gomel, a small town a little further from Chernobyl than Kiev but to the north. Very high levels of dicentrics and ring chromosomes were found, although translocations were not sought. These aberrations are intrinsically unstable and could cause long-term genetic defects. Figure 6.9 shows a human dicentric chromosome induced using X-rays *in vitro*.

Although some effects may not be known for a very long time, perhaps even for several generations, because the results may be quite subtle, some changes are already becoming apparent.

So far no increase in childhood leukaemias has been found in the vicin-

ity of Chernobyl, so presumably there has been no increase in leukaemias in nearby sovereign states. However, this does not mean that such an increase will not occur in future. Generally speaking if no effect is seen at high exposure then low-exposure areas should be relatively safe. This does not include areas where exposure was so high that neoplastic cells could not survive.

A more disturbing result is the increase in thyroid cancer in children from the Chernobyl area, northern Ukraine (276 cases), southern Belarus (251 cases) and part of the Russian Federation. The rate in some areas has reached 100 cases per million children per year. The basal rate of thyroid cancer incidence is between 0.5 and 3.0 per million per year.

It is known that the mutation rate, as measured by the number of translocations, varies from tissue to tissue. To determine whether the translocation rate also varies between species it was measured in three different species of animal under identical conditions. By measuring the translocation yield at the first meiotic division after exposure to 100 rad radiation it was found that 7.7×10^{-4} translocations per rad per cell occurred in humans and marmosets whereas mouse was far more robust with 2×10^{-4} translocations per rad per cell.

In somatic cells with a constant turnover, therapeutic radiation exposure has revealed that, while the number of translocations is more or less constant over a period of years, the numbers of rings and dicentrics is reduced. This change probably reflects the stability of most translocations while other rearrangements are less stable during cell division.

Translocations involved in Burkitt's lymphoma are shown in Figure 6.10. This is a good indicator of the alterations which are required to induce proliferative changes. An oncogene needs to be switched on in some way, such as translocating the oncogene to a gene that is constantly being transcribed. The classical translocation in leukaemias creates a small chromosome that is the product of a translocation between chromosomes 9 and 22. This derived chromosome is unusual in having a name; it is called the Philadelphia chromosome. It is the consistency of well-characterized translocations within specific

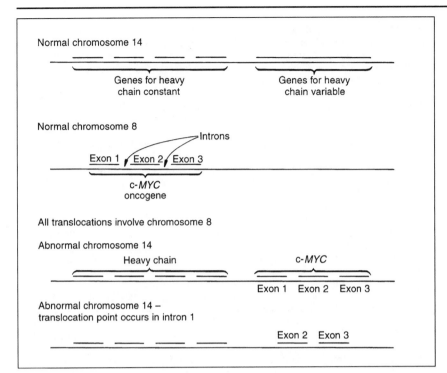

Figure 6.10 Translocations associated with Burkitt's lymphona. The translocation of the c-*myc* oncogene from chromosome 8 to chromosome 14 results in c-*myc* becoming closely associated with a highly transcribed heavy-chain immunoglobin and therefore becoming transcribed itself.

cancers that has prompted the detailed study of the underlying molecular lesions (Box 6.5).

Consistent chromosomal changes have been noted in radiation-induced leukaemias in mice. But this does not mean that the translocations and deletions induced by mutagens are also consistent. Most chromosomal damage is of little consequence to the cell, or else causes premature cell death. In the most studied of experimental animals, humans, the translocations involved in leukaemias have been looked at in the most detail. These haematological disorders originate from a diverse range of translocations, frequently involving highly transcribed genes, such as immunoglobins or T-cell receptors. The precise breakpoints and rearrangements are extremely varied, with degenerative changes also occurring with time, which has historically made the elucidation of the important translocation quite difficult. Not shown in Table 6.6 is the large number of rearrangements which involve chromosome 11, usually at q23. Some of these are found in lymphoblastic leukaemias [t(1;11), t(4;11), t(11;19)], whereas others occur in myeloid leukaemias [t(1;11), t(2;11), t(6;11), t(9;11), t(10;11), t(11;17), t(X;11)]. Detailed work on cloned cells with these breakpoints has shown that they are associated with a gene that is homologous to the trithorax gene of *Drosophila*, which is a developmental regulator.

Box 6.5   Control of deliberate translocations: malaria

Throughout the world more than 300 million people are suffering from malaria at any one time. Approximately 1 million people die each year from malaria, of whom 90% are in Africa. The disease is caused by the intracellular parasite *Plasmodium falciparum*. *Plasmodium* has one of the most complex and highly evolved life cycles of any protozoan parasite. Two hosts are involved: anopheline mosquitoes and vertebrates. In mosquitoes there is a sporogonic cycle and in humans two schizogonic cycles. All stages except that in which the parasite grows in the gut wall of the *Anopheles* mosquito are haploid.

Like trypanosomes, *Plasmodium* manages to evade the host immune system. This can only be achieved by changing its antigenic coat. *Plasmodium* has developed a way of effecting antigenic change at the rate of 2% per generation. Just as in trypanosomes it has been demonstrated that variation is associated with the chromosome telomeres. It is quite likely that this process is associated with all of the 14 chromosomes found in *Plasmodium*. Shuffling the variable antigen genes enables surface antigen expression to be controlled by genes associated with telomeres. *Plasmodium* telomeres contain repeat sequences, of which the repeat motif is TTTAGGG. This is so similar to other telomere consensus sequences that it can be thought of as a common denominator that is essential to normal telomere activity.

The translocation breakpoints associated with *Plasmodium* telomeres do not appear to be distributed in a random fashion but are associated with the eight or so base pairs that form the linker between the nucleosomes. Nucleosomes are made up of histone proteins and about 146 base pairs of DNA which is wrapped around them. This nucleosome structure is remarkably consistent throughout eukaryotes.

Since not all internucleosomal DNA is suitable for breakage, we can surmise that there is another, unknown, controlling process involved.

A similar situation is found in the chromosomal rearrangements of solid tumours. Historically these have not been so well characterized as the translocations of haematological disorders. There are a number of reasons for this: long-term culture causes degenerative changes which can mask the true nature of the mutation event and the very wide range of solid tumours in comparison with leukaemias, some of which are very rare, has made the build-up of data difficult. Among the consistent changes that have been recorded in solid tumours are complex rearrangements, double minutes (especially in neurogenic neoplasms), deletions, isochromosomes and acentric fragments. It is only recently, with the advent of *in situ* hybridization, that it has become possible to determine the origin of many of these marker chromosomes because their rearrangements make them virtually unrecognizable by traditional cytogenetic techniques.

6.7   Fragile sites

Although many chromosomes are known to have sites of fragility, there are few in which the nature of the process is clearly understood. Most work on fragile sites has been done using human material, and most of this work has been associated with the X chromosome. The chromosomal mutations associated with fragile sites are particularly interesting, as is the resultant phenotype.

Fragile X syndrome [Fra (X)] is the second most common form of inherited mental retardation after Down's syndrome. Figure 6.11 shows the appearance of the fragile X site when it is expressed in culture.

Clinical features of the disease, apart from mental retardation, include macro-orchidism, loose joints and a typical facial appearance including prominent ears and lower jaw, long face and high forehead. In the general population approximately 1 in 1250 males and 1 in 2000 females are affected. In the past, diagnosis has been made cytogenetically by the detection of a folate-sensitive fragile site at Xq27.3 (FRAX A). To expect such a simple test to detect all cases was perhaps overambitious as an increasing number of molecular variants have been discovered.

In 1991, the gene, designated *FMR-1* (fragile X mental retardation 1), which was thought at that time to be exclusively responsible for the condition, was characterized. Since that time it has become apparent that the area around the *FMR-1* gene is also responsible for disease expression. Factors involved include the size of a trinucleotide tandem repeat of CGG which is present in one of the exons of *FMR-1* and the methylation status of a CpG island 250 bp upstream of the trinucleotide repeat (for more detail on CpG islands, see Chapter 3). Any abnormal expression of these factors results in fragile X syndrome.

Figure 6.11 Fragile X expression in humans. (a) Solid-stained preparation of human chromosomes. The fragile site is marked. To be certain that this really is an X chromosome it has to be destained and banded. (b) The same metaphase as in (a) but having been G-banded. The X chromosome is marked, although the very small fragile fragment has been completely digested. These two images also serve well to demonstrate the structural changes which take place with the action of proteolytic enzymes on chromosomes. Courtesy of S.C. Rooney.

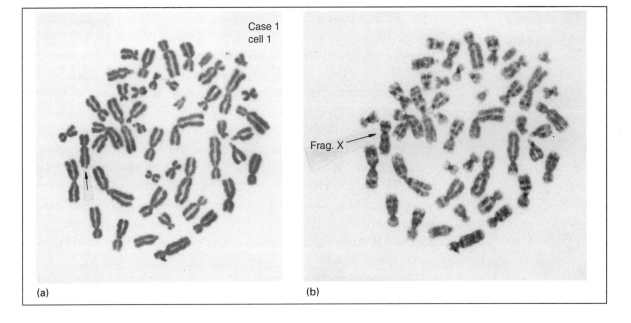

Case 1
cell 1

Frag. X

(a)   (b)

It is the expansion of the trinucleotide repeat that is presumed to be responsible for the folate sensitivity of the chromosome. A clinically normal person has between 6 and 50 of these repeats. There are two mutation states, the first of which is a premutation with between 52 and 200 trinucleotide repeats. An individual with this premutation is symptom free and has carrier status (30%) carrier females and 20% of non-transmitting males). The full mutation is described as having more than 600 repeats, although repeats may number in excess of 1000 and are associated with methylation of the CpG island. Individuals with the full mutation show expression of fragile X syndrome; all males and 50% of females are mentally retarded to a varying degree.

The premutation is unstable and can progress to the full mutation. Although there is some doubt as to the exact mechanism underlying the changes, it is known that it only takes place when the FRA(X) chromosome is transmitted through the maternal line. The risk of this change increases exponentially with an increase in the number of repeats present in the premutation of the maternal carrier chromosome. As can be readily appreciated, this is a very unusual method of inheritance and the risk of mental impairment can be anticipated; this is known as the Sherman paradox and is manifested as an increase in the proportion of mentally retarded individuals in successive generations of an affected family. Interest in repeat expansion has been fuelled by its implications for other diseases, such as myotonic dystrophy.

An interesting point about FRA(X) patients is that germline and somatic cells are not identical. Sperm may not have the complete mutation that is present in peripheral lymphocytes. It is suggested that change from premutation to full mutation is post-zygotic and does not occur in meiosis. This idea is supported by the observation that affected individuals are mosaic for repeat numbers. It may be that expansion takes place during a particular window of embryonic development.

While trinucleotide repeat expansion is believed to cause cytogenetic expression of fragile X, methylation of the CpG island is strongly correlated with absence of *FMR-1* mRNA. Fetal studies have shown that *FMR-1* mRNA is present in chorionic material in which methylation is not established, but not in fetal tissues where CpG is methylated. Methylation is not immediate and takes place around week 10 of gestation (for more details on gene methylation and its controlling role see Chapters 5 and 8).

This syndrome is also expressed in cases which cannot be detected cytogenetically. In these cases point mutations and deletions of the

FMR-1 gene seem to be responsible. The *FMR-1* protein seems to be highly conserved across species, but does vary between tissues.

6.8 Transposable elements

Transposable elements can cause mutagenic changes in a genome, but their mode of action is very different to the mutagens described above. Transposable elements were first discovered over 40 years ago by Barbara McClintock when she was studying unstable colour mutation in maize kernels. This was designated the Ac/Ds (activator/dissociator) system. Since then many more transposable elements have been discovered.

Transposable elements are mobile genetic elements that are able to insert at different positions in the genome without requiring homologous DNA sequences for recombination. They can affect the function of the genes with which they become associated by direct insertion into either the coding, promoter or control sequences, producing mutant alleles. They can be distinguished from a 'normal' mutation because, although they reduce or completely prevent the expression of a gene, they frequently revert to a non-mutant wild type, producing a mosaic pattern of expression for the mutated genes. For example, if the transposable element inserts into a colour gene, a variegated phenotype would be produced as a result of the random excision events (Figure 6.12).

Figure 6.12 The pallida gene of *Antirrhinum majus* encodes a product required for the synthesis of red flower pigment. Left: a flower from the progenitor containing the Tam3 transposon. The next three rows show variants of stable integration that confer flower colour. The final row shows the spotted phenotype due to transposition events. Reproduced with permission from Coen *et al.* (1986), *Cell*, **47**, 286, Figure 1, copyright Cell Press.

These events are possible because the transposable elements themselves contain the genes coding for enzymes which induce duplication, excision, integration, etc. The enzymes coded for depend on the nature of the element and its mode of transposition. Some of the necessary enzymes can be 'hijacked' from the cell's own DNA replication and repair systems. It should be noted that transposition is not restricted to the nuclear genome, but also affects chloroplasts and mitochondria.

It is thought that transposable elements have evolved from sequences already present within the cell, or have been acquired from outside, as a result of events such as viral infections. Their distribution appears to be ubiquitous within all types of organisms and at all levels.

They can be divided into two main types depending on the nature of their mobility.

6.8.1 Transposons

If transposable elements replicate via a DNA intermediate, they are called transposons. They carry the coding sequences for the transposase enzyme and DNA sequences which are recognized by the transposase as necessary for transposition (the inverted repeat sequences at the ends of the transposon). It is thought that the transposase enzyme recognizes these sequences, bringing them together to allow the complementary sequences to interact, and introduces staggered nicks at the ends to induce excision. The excision and integration process is then essentially a 'cut and paste' procedure within the genome, i.e. the original transposon moves; duplicates are not made.

The excision and integration events are often imperfect events producing small deletions, duplications or inversions of the end repeat units of the transposon. These are characteristic for each species and are often called footprints. These can be used to determine the extent of transposon movement within an organism. For example the Ac system leaves behind an 8-bp and the En/Spm a 3-bp footprint.

Non-autonomous versions can occur which carry only the recognition sequences and rely on a *trans*-coded transposase. The best known of these is the Ac/Ds system in maize, which as it contains two elements is known as a controlling element system. Ac can promote its own transposition or that of Ds to another site within the genome, but Ds cannot move unless Ac is present in the same cell. This is because the Ds elements are derived from Ac, but with internal deletions which knock out the transposase enzyme function. Another example of this type of system is En/Spm, also in maize.

6.8.2 Retrotransposons

If replication occurs via an RNA intermediate the transposable elements are called retrotransposons. It is thought that retrotransposons move by producing an RNA copy, which is then reverse transcribed into DNA and subsequently integrates into the genome in a random position. These represent a more diverse group than those that reproduce via a DNA intermediate.

Retrotransposons can be subdivided into two sets.

1. Viral. These are related to the retroviruses. Originally they all encoded their own reverse transcriptase, RNA-binding activity and

Table 6.9 Types of transposon

Name	Host species	Size (kb)	Number of copies	Structure of termini
Transposons				
Ac	Maize	4.6	Variable	10-bp inverted repeat
Ds	Maize	Less than 4.6	Variable	10-bp inverted repeat
En	Maize	8.3	Variable	13-bp inverted repeat
Spm	Maize	Less than 8.3	Variable	13-bp inverted repeat
Mu1	Maize	1.4	Variable	200-bp inverted repeats
P	*Drosophila*	2.1	Variable	31-bp inverted repeat
Tc1	*Caenorhabditis*	1.6	20–200	50-bp inverted repeat
FB elements	*Drosophila*	0.1–20	2000–4000	Short inverted terminal repeats of variable number
Tu	Sea urchin	0.1–20	200–400	840-bp inverted repeats
Tam-1	*Antirrhinum*	17	100 +	13-bp inverted repeat
Retrotransposons (viral)				
Ta-1	*Arabidopsis thaliana*	5.2	1–3	514-bp long terminal repeats
Tnt-1	Tobacco	5.3	100 +	600-bp long terminal repeats
Tst-1	Potato	5.1	1	283-bp long terminal repeats
del-1	Lily	9.3	13 000 +	2.4-kb long terminal repeats
IFG-7	Pine	5.9	10 000	333-bp long terminal repeats
Cin-1	Maize	0.7	1000	690-bp long terminal repeats
Bs-1	Maize	3.2	2–3	302-bp long terminal repeats
Wis-2	Wheat	8.6	200	1.7-kb long terminal repeats
PDR-1	Pea	5	50	156-bp long terminal repeats
BARE-1	Barley	8.5	Multiple	1.7-kb long terminal repeats
Ty family	Yeast	5–7	5–35	330-bp long terminal repeats
Copia-like family	*Drosophila*	5–7	20–100	266–512-bp long terminal repeats
Retrotransposons (non-viral)				
Cin-4	Maize	6.8 +	25–50	Oligo-A track plus small direct duplications
Del-2	Lily	4.5	240 000	
I	*Drosophila*	5.3	Multiple	Poly-A tail
LINES	Mammals	3–7	10 000–100 000	
SINES	Mammals	0.3	10 000–100 000	

the means for integrating into genomes. They contain LTRs (long terminal repeats) ending in 5′ TG … CA 3′.

2.   Non-viral. These are retropseudogenes often derived from RNA genes within the cell. For example, the human Alu family of sequences, SINES (short interspersed repeat family) in mammals are derived from 7SL RNA. Other family members are derived

---

Box 6.6   Transposon tagging

Transposons in the past have (by accident) allowed the successful isolation of genes normally difficult to access such as the homeotic gene *lin-12* in *C. elegans* and the *Drosophila* DNA polymerase gene (RpII).

A particular aim is to isolate developmental genes in plants, for example those involved in the control of flowering in response to daylight and temperature, determination of flower shape, etc. These are genes which are difficult to isolate because very little is known of their protein product, the pathways involved, tissue specificity or timing of expres-

sion. Insertion of a transposon into these genes produces a characterizable mutant phenotype, often with variegated expression patterns. The gene involved can be isolated by using the transposon itself as a probe (or 'tag') (Figure 6.13).

So far they have mainly been used to isolate genes involved in the anthocyanin (colour pigment) pathways of maize and snapdragon as these are easy to detect. *Cf-9*, a fungal resistance gene in tomato, has recently been successfully isolated using the Ds element of maize.

The system which has been

adapted for general use is the Ac system from maize. This is because it is active in several species of plant (tobacco, potato, tomato, carrot, *Arabidopsis* and petunia), is the best-understood system, has a simple structure and also transposes at higher frequencies than other types tested. Alterations have been made to the original structure to increase the frequency of transposition and improve the targeting of transposition events, which at the moment are still largely random.

Figure 6.13 Procedure for isolating genes via transposon tagging.

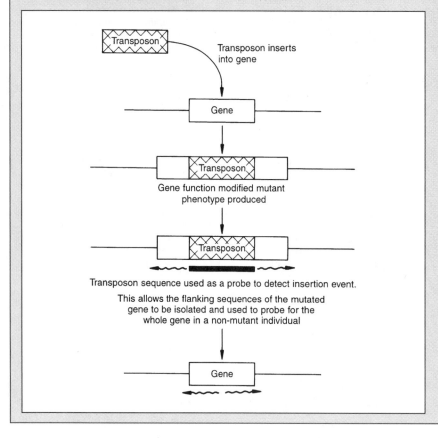

from RNA polymerase III and tRNA genes. The LINES (long interspersed repeat family) in mammals contain an open reading frame with homology to retroviral reverse transcription genes.

Although many new transposons are continually being identified (Table 6.9), often by sequence analysis, such as sequence conservation of the binding site for reverse transcription or analysis of dispersed repeats, very few are active; most are non-functional, inactivated by deletions and mutation. However, the ability of a transposon to move can only be inferred from the products of mutation or by comparing sequences of regions of DNA from different strains of the same organism and showing the insertion of DNA. It has been shown that they can be activated by stress such as heat shock, UV irradiation or cell culture.

Apart from the direct mutation function described here, transposable elements are thought to be responsible for a wide variety of biological phenomena including drug resistance genes in micro-organisms, antigenic variety in trypanosomes, hybrid dysgenesis in *Drosophila* and possible antibody diversity in man.

They are also thought to play a role in evolution by enriching the number of repetitive sequences present within the genome and it is becoming increasingly clear that they make up a large proportion of dispersed repeat families in the genome, the form and distribution of which must have had and probably is continuing to play, a major role in genome organization and function. It is also thought that they provide the genome plasticity needed for adaptation to unusual situations. This would be particularly important in the case of plants, which cannot move to escape stress. Their maintenance in the face of mild selective pressure must be due to occasional transposition and their potentially beneficial role.

While transposons are responsible for a certain amount of mutation, they are now being actively manipulated as cloning vehicles, particularly for plants (Box 6.6).

7 Chromosome mapping

Summary

Chromosome mapping is not just about localizing important genes, it is also about the nature, organization and interaction of sequences on a chromosome. The relationship between the repetitive and single-copy fractions and between heterochromatin and specialized chromosome structures. It also provides the key to understanding the nature of genes and their action. For example, there is great interest in why some cancers are associated with certain translocations, whether this represents the disruption of a vital gene action or the bringing together and altered expression of two genes normally kept far apart in the genome. This type of question can only be answered by mapping studies of the genes involved and the surrounding control regions.

Mapping can be achieved at several levels: single chromosome, gene cluster and gene level, and ultimately with the full DNA sequence of the promoter, regulator, coding and non-coding regions of the gene under investigation. The difference in size between a single chromosome and a DNA sequence is immense in molecular terms. Hence, mapping tends to be a layered approach, the resolution limited by the techniques employed. Figure 7.1 shows the resolution boundaries of the techniques to be described later in the chapter.

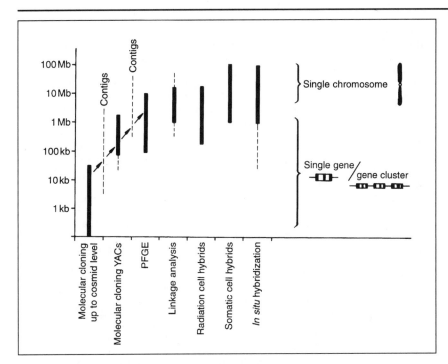

Figure 7.1 Resolution constraints of mapping techniques.

7.1 Introduction

The early part of the book has covered sections on chromosome structure and identification, but how does this physical structure relate to a DNA sequence? The increased use of molecular biology and the recent initiation of the Human Genome Project (Box 7.1) has focused attention on the need to isolate specific genes as well as the need to localize them to a chromosomal region and determine their immediate molecular environment.

The first point to note is that there are two types of map: genetic and cytological.

1. Genetic (often called linkage maps). Genetic maps identify the linear arrangement of genes on a chromosome and are assembled from meiotic recombination data. These are theoretical maps which give the order of genes on a chromosome. They cannot pinpoint the physical whereabouts of genes or determine how far apart they are. Distances on this map are not directly equivalent to physical distances. The unit of measurement is the centimorgan (cM).

2. Cytological (or physical maps). Cytological maps identify the actual physical position of genes on a chromosome. Distances are

---

Box 7.1   Human Genome Project

The primary aim of the Human Genome Project (HGP) is to sequence the entire human genome by the year 2005. The scale of international cooperation required is unique in the field of biology. Although the idea for such a project originated in 1984, financing was not achieved until 1988. In 1990 HUGO (Human Genome Organization) was formed to coordinate the efforts of all countries involved in the HGP. Involvement in the project encourages international pooling of resources, collaboration and free exchange of data.

Goals and time scales have been set. For example, the first 5-year plan, ending in 1995, aimed to produce a high-resolution linkage map with an average spacing of 2 cM and a physical map consisting mainly of YACs, with a sequence tagged site every 300 kb (Section 7.2.4). This is on target. The plan for the next 5 years is the identification and localization of genes to produce a transcription map.

There is also interest in the role of physical spacing in gene regulation, the organization of repetitive elements and non-transcribed spacers, the distribution of transposons, pathologies of DNA including mutations and rearrangements, the sequencing and characterization of other model genomes, etc. This intense body of work has led to the rapid improvement in automated sequencing techniques and database handling. The output is vast and the time scales for target achievements are decreasing all the time.

This project is not without controversy. The potential exists to obtain detailed knowledge of individuals' genetic make-up, including predisposition to disease. This has led to discussions about the need for privacy of results and whether insurance companies or employers ought to be told of any 'faults' in their clients' or employees' DNA that would affect life expectancy, insurance claims, work performance, etc., indeed whether they ought to insist on DNA tests before taking anyone on. There has been much recent press coverage regarding genetic tests for Huntington's chorea. This is a late-onset disease and affected individuals must endure relatively long periods of suffering before what would be considered an early death. Because of the late onset, couples have generally started a family before discovering that one of them is affected. Closely related family members are offered genetic testing and counselling. If someone tests positive for the disease, apart from the obvious trauma, this result will affect their whole life as regards potential quality and expectations. If insurance companies learnt of such a result, their attitude to pensions, medical insurance and life cover for that person would dramatically alter. People in this situation obviously want to make financial provision for their families if they have not done so already, but would suddenly find that the traditional institutions are not interested in covering them, a similar situation as has happened with AIDS patients. With the increasing number of genetic tests available, this dilemma will only increase.

---

measured in base pairs (bp), kb kilobases, (1000 bp, 1 kb) or megabases (1 000 000 bp, 1 mb).

How each type of map is obtained will now be discussed.

7.2   *Genetic maps*

The procedure of genetic mapping is often called linkage analysis and is based on Mendel's second law, the law of independent assortment, which states that the members of different pairs of alleles assort independently of each other when the germ cell is formed. This is only true if the genes are on separate chromosomes. If they are on the same chromosome, they are said to be linked. The frequency of recombination of a pair of linked genes is, under standardized environmental conditions, constant and characteristic for that pair of genes.

In 1911, T.H. Morgan hypothesized that the strength of linkage of a pair of genes, i.e. the amount of linkage and recombination observed, is a function of the distance on the chromosome between the genes in question. The greater this distance, the more likely, in general, that recombination will occur between these genes.

These data can only be obtained by crossing organisms with quantifiable segregating differences and analysing the offspring of such a cross for each of the parental characteristics under study. The frequency of recombinants (i.e. those which show a mix of both parental types) is a measure of whether the genes are linked or not and, if so, how close together they are on the chromosome (Figure 7.2).

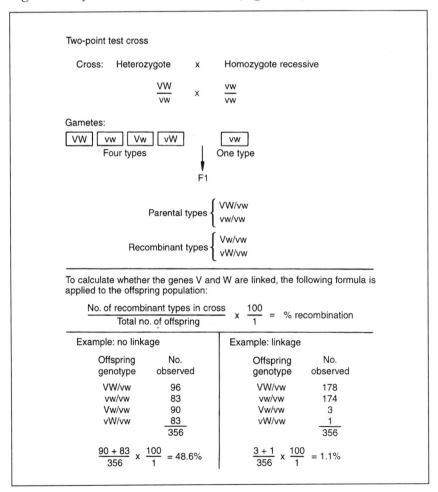

Figure 7.2 Two-point test cross.

In the example given, a heterozygote for dominant genes V and W is crossed with a homozygous recessive. The heterozygote would be expected to produce four types of gametes via independent assortment. The homozygote would produce only one type. Hence, when these combine, because of the recessive nature of the genes supplied by the homozygote, the offspring show the phenotype of the genes supplied by the heterozygote, i.e. four different genotypes. The number of each type is a measure of whether the genes are on the same chromosome. If there is no linkage (i.e. V and W are on different chromosomes) the ratio of parental types to recombinants would be approximately 50:50. With linked genes (i.e. present on the same chromosome) this ratio could be significantly different. The percentage recombination is generally expressed as centimorgans (cM) (1% recombination equals 1 cM).

If genes are linked, the amount of linkage (or recombination) is a measure of how far apart the genes are on the chromosome. The percentage recombination values are approximately additive, and therefore it is possible to expand this technique using different crosses to determine the order of a series of genes and to build up a map based on these measurements.

This is simply demonstrated in the following example, to determine the order of genes X, Y and Z (a three-point test cross, Figure 7.3).

In this basic example, the figures do not exactly fit together; no account has been taken of the possibility of double crossovers between the three alleles or interference. In 'real' crosses this can be compensated for to a certain extent using set mathematical formula, such as Kosambi's mapping function, producing very accurate genetic maps with extensive collections of markers as shown in Figure 7.4.

The result of multiple crosses of different genes is a linkage map. In the initial stages of genetic mapping, the chromosomal assignment is generally unknown, so a series of gene groups emerge which have been proven to be linked to each other (linkage groups). As the genetic mapping expands, the number of linkage groups equals the haploid number of chromosomes. Physical mapping can then allocate these groups to specific chromosomes.

Genetic maps rely on the presence of quantifiable differences between the individuals of a cross and on the ability to differentiate between and analyse the offspring. They can be constructed using both morphological and molecular markers (e.g. isozymes, restriction fragment length polymorphisms, randomly amplified polymorphic DNAs, sequence tagged sites).

Figure 7.3 Three-point test cross.

Three point test cross

$$\frac{XYZ}{xyz} \quad x \quad \frac{xyz}{xyz}$$

Gametes: | XYZ | xyz | | xyz |
| XyZ | xYZ | One type
| Xyz | xYz |
| xyZ | XYz |

Eight types

Offspring

Genotypes	No. observed	Recombinant between X and Y	Recombinant between X and Z	Recombinant between Y and Z
XYZ/xyz	853	–	–	–
xyz/xyz	926	–	–	–
Xyz/xyz	51	Yes	Yes	–
xYZ/xyz	42	Yes	Yes	–
xYz/xyz	52	Yes	–	Yes
XyZ/xyz	60	Yes	–	Yes
xyZ/xyz	7	–	Yes	Yes
XYz/xyz	9	–	Yes	Yes

2000

Recombination between X and Y

$$\frac{51 + 42 + 52 + 60}{2000} \times \frac{100}{1} = 10.25\%$$

Recombination between X and Z

$$\frac{51 + 42 + 7 + 9}{2000} \times \frac{100}{1} = 5.45\%$$

Recombination between Y and Z

$$\frac{52 + 60 + 7 + 9}{2000} \times \frac{100}{1} = 6.4\%$$

Map   Gene order

X   5.45   Z   6.4   Y

7.2.1 Morphological markers

Morphological markers are physical characteristics, such as round or wrinkled pea seeds (as in Mendel's experiments), which can be counted directly. This may be considered a great advantage at first, but their number is very limited and most maps are largely composed of molecular markers.

7.2.2 Molecular markers

Molecular markers are generally phenotype neutral in effect. As their detection is at the molecular level, most molecular markers behave in a

Figure 7.4 RFLP map of chromosome 1 of tomato. Reproduced from Tanksley *et al.* (1992), *Genetics*, **132**, 1146, Figure 1.

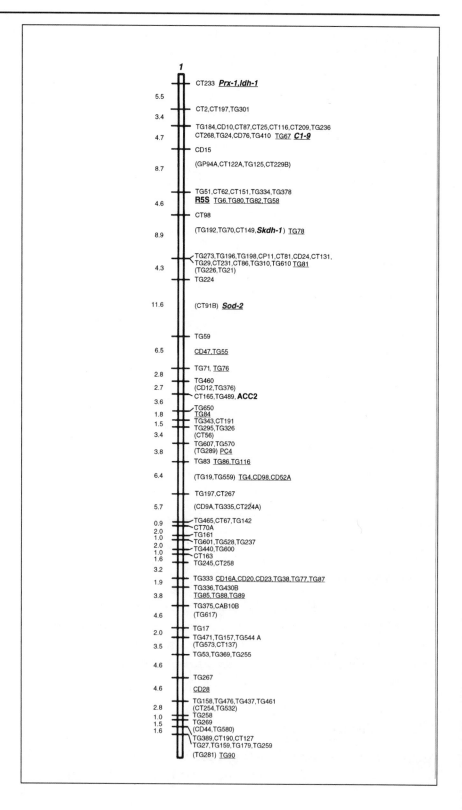

co-dominant fashion, ie. all alleles can be detected, unlike morphological markers, which are subject to dominant–recessive and epistatic interactions. Some of the more common types of molecular marker are described below.

Isozymes

If a population contains different forms of an isozyme (enzyme), these can be extracted and visualized on special polyacrylamide gels. Isozyme maps have been published for several plants, including tomato, maize, wheat and pine. However, the number of isozymes available tends to be fairly limited.

Restriction fragment length polymorphisms (RFLPs)

RFLP sequences are short pieces of random DNA which are non-coding and neutral in effect, present in single to low copy number, isolated from either the organism under study or a closely related species. They have no ascribed function and are generally uncharacterized. They are detected on Southern blots (Box 7.2).

RFLPs can be isolated in large numbers, and their primary use is to saturate the genome with markers which can be used as signposts to narrow down the position of genetically important sequences. Linkage of an RFLP with a physical characteristic or disease is often the first indication of a position for the latter.

There is now a complete genetic linkage map of the human and mouse genomes based on RFLP markers, with more being added all the time. In agronomy there are maps available for most of the important crops, such as tomato, wheat, barley and potato. Some crops, such as tomato and potato, have the same evolutionary origin (*Solanaceae* family) and therefore have the advantage that RFLPs isolated from one can be used directly on the other (Figure 7.6).

Random amplification of polymorphic DNA

This is a PCR (polymerase chain reaction)-based technique (Box 7.3).

A single short oligonucleotide (usually 10 bp) of randomly chosen DNA sequence (primer) is mixed with genomic DNA and subjected to PCR with a very low annealing temperature (37°C, compared with the more usual 55°C). The primers will anneal to the genomic DNA at random and prime the amplification of several DNA fragments. The

Box 7.2　Southern blot/RFLPs

How are RFLPs produced (Figure 7.5)?

The genomic DNA of the organism under study is cut into pieces with restriction enzymes. This DNA is loaded onto an agarose gel and under the force of an electric current migrates down the gel according to size, the smallest pieces moving the fastest. The gel is denatured (i.e. the DNA made single-stranded) and blotted so that the DNA is transferred onto a solid nylon membrane. This membrane now contains single-stranded DNA sorted according to size (Southern blot).

An RFLP probe (which is also single-stranded) is labelled with radioactivity and incubated in solution (hybridized) with the blot. After washing, only homologous sequences will pair and remain attached to the blot. This blot is overlaid with photographic film and left in the dark. The radioactivity in the probe causes any region of the film in contact with it to become exposed. On developing, a pattern will appear on the film according to the hybridization pattern of the probe. Any polymorphisms in pattern (DNA) can be related to differences between the samples on the blot and correlated with the inheritance patterns under study.

Addition, deletion and rearrangement of restriction sites creates polymorphisms within populations. The profile of a series of polymorphisms can define the genetic make-up of an individual. Specific polymorphisms can sometimes be correlated with the pattern of inheritance of a disease, thus allowing a disease to be pinpointed to a certain chromosomal region. If tightly linked, the RFLP may provide the means for a test which can predict the occurrence of, or confirm the diagnosis of, a genetic disease.

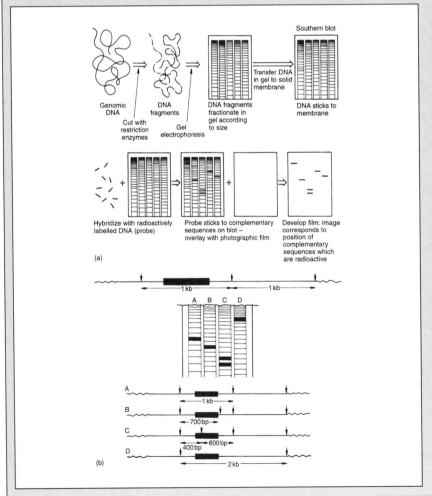

(a)

(b)

Figure 7.5 (a) Production of RFLPs. (b) Top: a stretch of genomic DNA; the arrows represent restriction enzyme sites and the coloured block the probe used to detect this region of sequence. The blot below this demonstrates how the sequence may vary between different individuals whose DNA has been cut using the same restriction enzyme. A, normal individual, probe hybridizes to a 1 kb region on the blot. B, mutation produces an additional extra restriction site in a region not covered by the probe. The probe hybridizes to a smaller piece of DNA. C, mutation producing an extra restriction site within the region covered by the probe. Hence the probe will then hybridize to two smaller regions of DNA. D, deletion of a restriction site. The probe hybridizes to a larger piece of DNA.

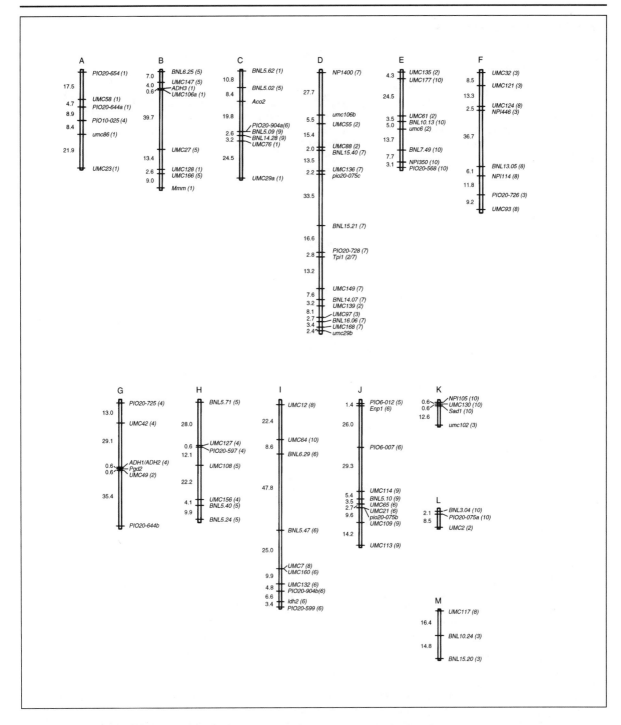

Figure 7.6   Linkage map of *Sorghum* chromosomes produced from maize RFLPs (both belong to the grass tribe Andropogoneae, albeit distantly related). Note that Sorghum has a haploid complement of 10, but there are 13 linkage groups. The small pieces K, L, M are probably portions of other groups, which so far have not been linked up. Bar distances are in centimorgans. Reproduced from Whitkus *et al.* (1992), *Genetics*, **132**, 1120, Figure 1.

Box 7.3 Polymerase chain reaction

Polymerase chain reaction (PCR) causes the amplification of a segment of DNA between two regions of known sequence. The reaction utilizes two different oligomers, which are short pieces of synthetically produced DNA (between 10 and 30 bp) that are complementary to known sequence on opposite strands of the DNA and contain free 3′ -OH groups, which allow sequence extension by DNA polymerase. The primers anneal to their complementary target sequences on denatured DNA and the region between them is amplified (polymerization) by a thermostable DNA polymerase. The cycle of denaturation, annealing of primers and polymerization of the DNA between is repeated many times. Each cycle results in a doubling of the amount of product and hence massive amplification of DNA sequence occurs. The products can be visualized on an agarose gel.

The DNA polymerase (*Taq* DNA polymerase) is produced from a thermophilic bacterium, *Thermus aquaticus*, which can survive repeated DNA denaturation steps at 95°C (most DNA polymerases act at 15°C and would be almost instantly destroyed).

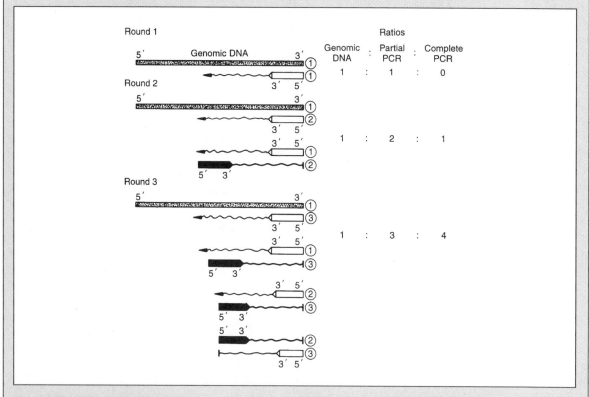

Figure 7.7 PCR amplification of a single strand of genomic DNA with two primers (primer 1, white, primer 2, black). The encircled numbers denote which round of PCR the stretches of DNA originated from. The same process will occur on the opposite strand (not shown). Round 1. Only one of the primers can bind to the genomic DNA (because of polarity) and produces an ill-defined amplification product of varying length (partial PCR, indicated by arrowhead). Round 2. The two products are denatured and are available for the second round of synthesis. Another primer 1 anneals to the genomic DNA and causes polymerization as in round 1. The previous primer 1 product is available for hybridization with primer 2; elongation using this primer will only continue until the 3′ end of primer 1, creating a defined product (complete PCR). Round 3. The four products are denatured and are available for the third round. Because of the nature of the previous products more complete products are produced compared with the previous round (1:3:4 or 50%, compared with 1:2:1 or 25%). In the following round this ratio becomes 1:4:11 (68%) and the next round 1:5:26 (81%). The resultant PCR solution contains a series of products: the genomic DNA, a defined product flanked by the two oligomer primers (complete) and a longer ill-defined product, the result of single primer amplification on genomic DNA (partial). The latter increases in a linear fashion and does not contribute significantly to the final products as the complete PCR product increases in an exponential fashion and overwhelms any of the other products.

The process of PCR is demonstrated in Figure 7.7.

Uses of PCR include detection of the DNA of pathogenic organisms in clinical samples; forensic identification; analysis of mutants; for example in diagnosing genetic diseases caused by additions and deletions of DNA sequence, resulting in different-sized PCR products between affected and non-affected patients; DNA sequencing and cloning of genes. The number of uses is continually increasing.

PCR has tended to be controversial, especially in the field of forensic examination. One of the problems is that it is a highly sensitive technique, able to amplify DNA from a single cell, so contamination of samples is a real danger, not just from mix-up of samples, but also contamination from the scientist carrying out the test.

'random' in the title is misleading, because the same primer will always prime the amplification of the same DNA segment of a particular organism, providing experimental conditions are kept the same. The products are short DNA fragments that can be directly visualized on ethidium bromide-stained agarose gels (Figure 7.8).

The nature of the fragments amplified is highly dependent on the primer sequence and the source of genomic DNA. The pattern of bands on the gel can be used on its own for linkage analysis or the bands can be extracted from the gel and used to probe back onto a Southern blot (i.e. used as RFLPs).

This method is gaining popularity, particularly in agriculture, as much smaller amounts of DNA are required for the technique and results are also obtained much faster than with RFLPs. The one drawback with RAPD is that heterozygotes are not detected; the map consists of dominant markers only. This is because DNA polymorphisms abolish primer binding sites, therefore absence of a band represents any alleles at that site which fail to bind.

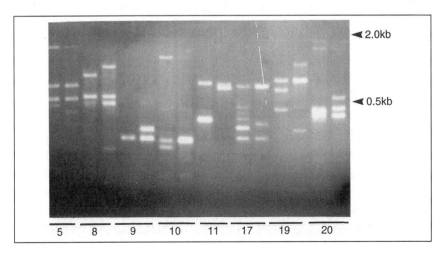

Figure 7.8 Example of RAPD gel from date palm DNA. The lanes are in pairs of male followed by female; the numbers denote the different primers used. There are many differences between the male and female samples as date palm is an outbreeder and has not been subjected to rigorous breeding procedures.

Sequence tagged sites (STSs)

These are short (200–500 bp) single-copy sequences that can be detected by PCR. Two oligodeoxynucleotide primers of approximately 20 bp each are manufactured; these bind to opposite strands and ends of the defined sequence. Traditional clones used in mapping, for example RFLPs, can be converted into STSs. The use of PCR speeds up the whole characterization process and is particularly useful for mapping YACs.

Microsatellites

These are also known as VNTR polymorphisms (variable number of tandem repeats). They are repeat sequences made up from a very short sequence motif, e.g. TG/CA. There are between 50 and 100 000 copies of such sequences interspersed throughout the human genome. They are highly variable with regard to copy number, and it is the copy number polymorphism that is used for mapping purposes. They can be assayed using PCR and the results viewed directly on agarose gels, which makes for a very quick test, or they can also be used as probes for Southern blotting.

7.2.3   Breeding manipulation

Breeding manipulation is usually required to produce genetic maps and is most accurate when inbred lines are used as the parental source material. This is because, if a particular characteristic is under study, for example disease resistance in cotton, it is much easier to cross two individuals which are as genetically similar as possible except for the characteristic under study (i.e. cross two inbred lines which differ only in their disease resistance characteristics). Any differences obtained in RFLP pattern will be much more likely to be due to the characteristic under study rather than individual-specific differences between the parental types.

Inbred lines are available for most of the common crop species. Animals tend to be more problematical, in that they may well object to being crossed for scientific purposes, produce fewer offspring (reducing the sample size) and have longer breeding cycles (reducing the number of potential crossings during the life of that organism). Most success has been achieved with mice, the development of laboratory inbred strains of which began many years ago. The other important organism is

Drosophila, which has a short life cycle and produces hundreds of offspring.

7.2.4   Linkage studies in humans

The linkage studies carried out in humans are all retrospective, based on family pedigree analysis and population studies. Because of the constraints on sample size, human linkage studies are subject to higher levels of error and hence pedigree analysis on its own has a relatively low resolution (1–5 Mb).

Reports have regularly been published since 1981 when Keats first produced a partial linkage map of nine human chromosomes (at a resolution of 16 cM). By 1987, the human linkage map, using RFLPs, covered the whole genome at a resolution of 10 cM. The latest of such maps is the result of collaboration between three major organizations and 110 collaborators, a truly international effort. The end result has 5840 markers placed at an average interval of 0.7 cM.

7.2.5   Quantitative trait loci

The construction of such maps has developed to such a level in some organisms that linkage analysis is no longer restricted to single genes. It is now possible to map polygenic traits. These are called quantitative trait loci (QTLs) (Box 7.4).

7.3   Cytological maps

This is the actual physical localization of sequences on chromosomes. The main techniques include: fluorescence *in situ* hybridization; flow-sorted chromosomes; somatic and radiation cell hybrids; dosage studies.

7.3.1   Fluorescence *in situ* hybridization (FISH)

FISH is rapidly becoming the most popular method for gene localization. The actual technique of *in situ* hybridization is described in detail in Chapter 3.

It is now possible to localize routinely single-copy genes on human metaphase chromosomes. In general, this produces a very crude map, as genes can only be localized to a specific band (resolution 1–5 Mb).

Box 7.4   Quantitative trait loci

QTLs define complex traits, i.e. any phenotype not exhibiting classic mendelian dominant or recessive inheritance attributable to a single locus, e.g. severe obesity, schizophrenia, fruit quality, etc. Requirements for detecting QTLs include a high-density molecular map, inbred lines (to minimize the individual-specific variation) and short generation times. Because of this their usefulness is limited by the type of organism that can be studied.

The progeny of a cross between two individuals differing in one or more characteristics are screened to determine their molecular marker map. A search is then made for correlations between segregating markers and the character of interest. Any associations should be due to linkage of the markers to the genes affecting the character. Because these traits are polygenic, analysis is highly statistical and linkage tends to cover a wide area, often a major portion of a chromosome arm. The result is a 'QTL likelihood map', which shows the most probable chromosomal position of the trait under examination in the form of a distribution curve superimposed over a map of the arm/chromosome. QTLs, because of their polygenic nature, are not necessarily restricted to single chromosomes: several may be involved, although not all to the same extent.

Traits successfully subjected to QTL analysis include hybrid vigour (maize), growth and fatness (pigs), hypertension in a rat model for primary hypertension in humans, epilepsy in mice and fruit mass, soluble solids, fruit pH in tomato.

Mice are now being used as animal models to analyse and dissect complex traits. The same mutation is bred into different genetic backgrounds (inbred strains) and the offspring of crosses examined for QTLs that affect the phenotypic expression of the mutation. For example, intestinal cancers induced by mutations in the mouse *Apc* gene have been shown to be subject to a modifier locus on chromosome 4. Type I diabetes in non-obese diabetic mice is affected by 12 separate loci. The results of dissection of such traits in mice can then be extrapolated to humans. The potential exists to refine our knowledge of the number of candidate genes involved in such diseases and, if the sequences are homologous between humans and mice, their cloning and characterization.

But this is a useful technique in the first step of determining the chromosomal position of a newly isolated gene. It is also used to check the cloned content of yeast artificial chromosomes to ensure that they come from one specific region of the genome and are not chimaeras (Figure 7.9).

For directly visualized mapping, i.e the ordering of genes along a chromosome, increased resolution can be obtained by hybridizing to extended prometaphase or surface-spread meiotic chromosomes, but the resolution remains in the region of 1 Mb. Recently, procedures have been developed to decondense and stretch the DNA in interphase nuclei, producing threads of chromatin. Hybridization to these can improve the resolution to 50–100 kb.

The number of different clones which can be detected at the same time by FISH is continually increasing. Each clone is labelled such that it has a unique colour under fluorescence; the limit at the moment is 12-colour FISH detection. Hence, it is now possible to select several genes from the same chromosome and hybridize and detect them all at once using different fluorochromes and detection methods. This means

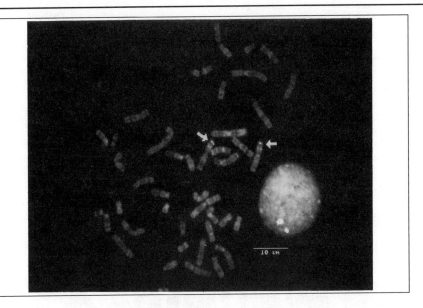

Figure 7.9 YAC clone of 36 kb hybridized to human chromosome 3p25–26. Courtesy of Dr C.G. Cee.

that they can be directly ordered along the chromosome, producing an image under the microscope like a string of coloured Christmas tree lights.

The use of routine single-copy FISH is mainly limited to humans and mice, as this is the current focus of research in the HGP. With plants the situation is more difficult in that the cell wall acts as a barrier to the immunological detection chemicals. Hence, very few examples of single-copy localization to plants have been published, and most maps in these cases are genetic.

Some organisms, such as *Drosophila*, possess polytene chromosomes. With these the *in situ* process is technically much easier. The 'single' copy sequence is duplicated hundreds of times within the polytene chromosome, vastly increasing the area for hybridization and detection. However, the number of organisms with usable polytene chromosomes is severely limited.

7.3.2 Flow-sorted chromosomes

Cultured cells are treated to stop cell division at metaphase. The cells are disrupted, releasing the chromosomes into a buffer. These are then stained with one or more DNA-specific fluorescent dyes such as ethidium bromide or propidium iodide, which fluoresce according to the amount of DNA present in each chromosome. Dyes such as Hoechst 33258 and chromomycin A3 are also used, which stain according to the base composition of the chromosome (A–T and G–C respectively).

Figure 7.10 Fluorescence intensity distribution of human male chromosomes stained with Hoechst 33258. Peaks have been assigned chromosome numbers. Reproduced with permission from Green (1990) Analysing and Sorting Human Chromosomes. *Journal of Microscopy*, **159**, 239, Figure 1.

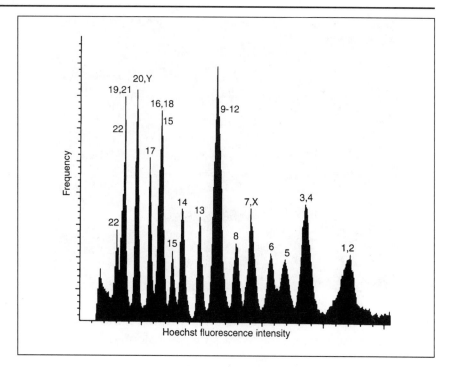

The stained chromosomes enter the sorter in solution (a 'chromosome' soup). The fluorescence intensities of the chromosomes are measured and analysed by a photomultiplier as they pass through two laser beams. The chromosomes then flow in a liquid stream in air to a point where this liquid breaks into droplets. These droplets are electrically charged and diverted according to this charge into different collection tubes. The charge added to each droplet depends on the results of the fluorescence analysis, which is in effect a measure of chromosome size. The result is many tubes containing differently sized chromosomes (Figure 7.10).

Thus, it is possible to obtain crude preparations of single chromosomes in each tube. This is relatively straightforward if all the chromosomes substantially differ in size, but problems arise with chromosomes of similar sizes, such as those in the human C group. With these it is possible to use human–rodent cell lines containing the human chromosomes required. Rodent chromosomes are significantly different in size and easy to differentiate from the human chromosomes present. Collections now exist for all 24 human chromosomes.

Flow-sorted chromosomes can be used as a source of chromosome-specific DNAs for *in situ* hybridization, DNA libraries and Southern blots. With the last mentioned, the chromosome extracts are sorted onto well-separated regions of nylon filters. A radioactively labelled sequence of unknown origin can be hybridized to this filter. The presence of a signal can then be correlated to a specific region of the filter and therefore a specific chromosome. This may be a faster process than doing direct localization using FISH, particularly in the case of cross-species hybridization, when experimental conditions make it more difficult to work on whole chromosomes than on Southern blots.

This chromosomal assignment can be narrowed down even further. Chromosomes can be flow sorted from cell lines containing rearrangements or from cells from patients with deletions and these used as a source of DNA for Southern blots.

Chromosome-specific DNA libraries can be used to analyse in depth the gene content of a particular chromosome much more easily than relying on isolating genes from the whole genome (for more information on DNA libraries, see Chapter 2). Also, chromosome-specific DNAs can be labelled for *in situ* hybridization and used to evaluate the chromosome content of somatic cell lines or used in cross-species evolutionary studies.

The flow-sorting process is more technically difficult to apply to plants owing to the obstructing presence of the cell wall and the presence of vacuoles, starch grains, plastids, etc. However, preliminary results have been achieved with tomato, separating the 12 chromosome pairs into three groups and three individual chromosomes. Tomato chromosomes are of a similar size, so different strategies may need to be applied to separate the chromosomes within the groups. A measure of success has also been achieved with *Vicia faba* (bean) chromosomes.

7.3.3 Somatic cell hybrids

Two somatic cell lines are established in culture. In the HGP, one of these lines is obviously human, the other rodent (usually mouse, hamster or rat). These two cell lines are fused in the presence of either inactivated viruses, e.g. Sendai virus or chemicals such as polyethylene glycol (PEG). The two genomes combine as the fused cells are cultured; the rodent cell line proliferates successfully, but the incorporated human chromosomes are randomly lost (Figure 7.11). This is initially a rapid process, which then slows down as only a few chromosomes remain, and this situation often remains stable.

The non-human chromosomes can be distinguished by their morphology and also staining properties. For example, when stained with Giemsa at an alkaline pH, mouse chromosomes stain bright purple and

Figure 7.11 Somatic cell hybrid panel formation. Two cell lines are fused. The resultant hybrid is grown up, but during this process loses human chromosomes at random. Stable lines with only one or two human chromosomes can enter a panel for gene mapping purposes.

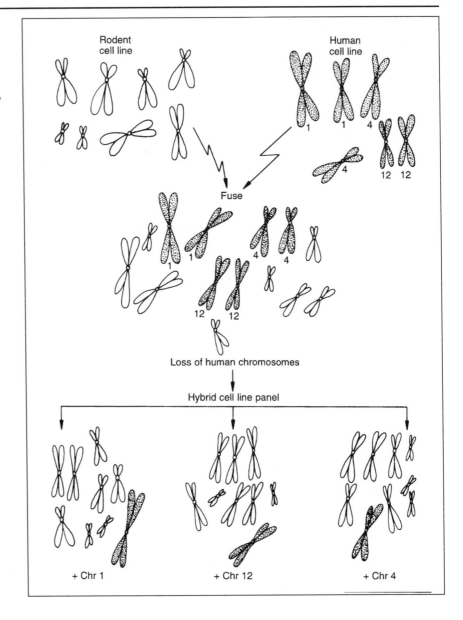

the human chromosomes pale blue. Cell lines can also be catalogued using banding techniques and *in situ* hybridization of human chromosome-specific DNA (chromosome painting).

Once the cell line has been defined in terms of human chromosome constitution, it can be used in a somatic cell hybrid panel. These contain numerous cell lines with different human chromosomes present in each. These are used as a source of DNA for Southern blots. A probe of unknown location is hybridized to the DNA of this panel and the

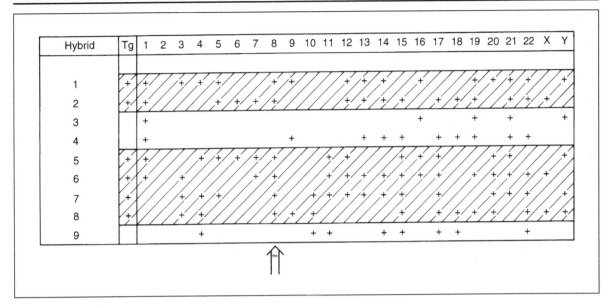

Figure 7.12 Results of a somatic cell hybrid panel hybridized with the thyroglobin (Tg) gene. The crosses denote the chromosomes present in each cell line. A positive signal for the Tg gene is correlated with the presence and absence of all chromosomes. Only chromosome 8 matches, in terms of distribution, the signal obtained with the Tg probe and therefore is the presumed location of the gene. Data taken from Baas *et al.* (1985) The Human Thyroglobin Gene: A polymorphic marker localised to C-MYC on chromosome 8 band q24. *Human Genetics*, **69**, 139, Table 1, copyright Springer-Verlag.

presence or absence of a signal from the probe can be correlated with the human chromosome constitution of the cell lines used (Figure 7.12).

Once the sequence has been limited to one chromosome, the process can be further refined to chromosomal regions. Human cell lines which have translocations or deletions already present or radiation cell hybrids are used.

7.3.4   Radiation cell hybrids

This is a variant on somatic cell hybrids and is often called irradiation and fusion gene transfer (IFGT). The initial hybrid human–rodent cell line contains only a single human chromosome with a full complement of rodent chromosomes and a selectable marker, such as neomycin antibiotic resistance. This cell line is irradiated, fragmenting all of the chromosomes. This irradiated cell line is recovered by fusing with another rodent cell line deficient for the selectable marker. Growth of the fusion products in selection medium allows for selection to operate in favour of fused hybrids against the background of non-fused rodent cells (Figure 7.13). Different cell lines are produced as some of the human chromosome fragments are integrated into these new cells. However, integration occurs in much smaller pieces than the original. The cell lines can then be examined for the presence or absence of molecular markers from that chromosome (using Southern blots, PCR of known markers or FISH).

Figure 7.13 Production of radiation cell hybrids. The donor somatic cell hybrid (rodent–human cell line containing a single human chromosome) is lethally irradiated. The chromosome fragments are fused to a recipient cell line. Selection produces a new rodent cell line containing fragments of the human chromosome and the original selectable marker.

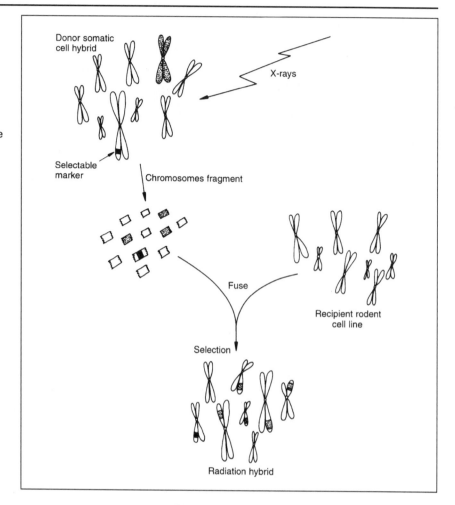

Figure 7.13 Production of radiation cell hybrids. The donor somatic cell hybrid (rodent–human cell line containing a single human chromosome) is lethally irradiated. The chromosome fragments are fused to a recipient cell line. Selection produces a new rodent cell line containing fragments of the human chromosome and the original selectable marker.

The theory behind this technique is that the further apart two markers are, the more likely that the irradiation will cause a chromosome break between them, placing the markers on separate chromosome fragments. By estimating the frequency of breakage, it is possible to determine gene order and estimate distance, building up a physical map of the chromosome. Measurements are presented in the form of centi-rays (cR). Conversion to kb depends on the initial radiation dose, e.g. 1 cR=30 kb at 6500 rad, whereas 1 cR=55 kb at 9000 rad.

The radiation dose is related to final chromosome fragment size, and therefore the dose can be tailored to the mapping requirements. For example, a dose of 5000 rad (50 Gy) to a cell line containing the human X chromosome produces at least 50% of fragments in the 2–3 Mb range (suitable for chromosome mapping). Increasing the dose to 25 000 rad (250 Gy) produces fewer than 6% of fragments in the

3 Mb range. These smaller fragments are more useful for positional cloning.

The most detailed radiation cell hybrid study so far has produced a full map of human chromosome 11, with 506 STS markers mapped at an average of 1 Mb.

7.3.5 Gene dosage

Gene dosage studies can be conducted on patients with additions and deletions for chromosomal regions. This can be shown at the DNA level using density measurements of the strength of radioactive signal found on Southern blots when the dosage of the gene in the patient is compared with that from a normal chromosome set. If both sets of the gene are present in the patient there will be 100% radioactive level on the blot when compared with normal. If the gene has been deleted from one chromosome, radioactivity levels will be down to 50%, and if an addition is present 150% activity. This enables a gene to be allocated to a specific region.

With humans, the application of this technique is limited by the amount of available material. It is, however, frequently used in plants, where aneuploid stocks for particular chromosomes are produced by traditional crossing methods, and is one of the main ways of mapping genes or linkage groups to plant chromosomes (Box 7.5).

Box 7.5 Aneuploid lines

These are plants which do not have the normal diploid (or euploid) chromosome number. The main types are: monosomics ($2n = 2x - 1$), which lack one chromosome; and trisomics ($2n = 2x + 1$), with an extra chromosome. Complete sets of monsomic lines are available for wheat, oats and tobacco. Complete trisomic lines are more numerous and include all the main crops such as wheat, oats, tomato, rice, rye and barley. They can be produced in a number of ways.

Primary trisomics are generally produced by crossing a diploid with a tetraploid. The result, in theory, is a triploid, but the extra set of chromosomes tends to be unstable and chromosomes are lost at random. The breeding stock may stabilize with the addition of only one or two extra chromosomes.

Monosomics can be produced using mutant plants in which pairing is defective (e.g. asynaptic wheat cv. Chinese Spring), resulting in aneuploid offspring. Certain tissue culture procedures can have a destabilizing influence on chromosome constitution. Regenerated plants may be aneuploid (see Chapter 10 for further details).

Just as with somatic cell hybrid lines, aneuploid lines can be produced containing only partial chromosomes with deletions of part or whole arms which allow mapping to be further refined from the whole chromosome level to that of a chromosome arm or segment.

While one of the main uses of aneuploid lines is chromosome mapping, alternatives include the production of alien addition and substitution lines for breeding purposes.

7.4   Correlation between genetic and cytological/physical maps

The two types of map and their method of production have been discussed. They both produce a gene order that is identical. However, genetic maps cannot be directly correlated with physical maps as there is no straight conversion factor from centimorgans to kilobases either within chromosomes of the same species or between chromosomes from different species.

Tomato probably has one of the most detailed classical linkage maps (Figure 7.14). It is also the subject of intense molecular analysis, which has since extended the classical map by at least 300 cM (20%).

Figure 7.14 Molecular linkage map of tomato chromosome 2. The RFLP linkage map is on the left, the classical (gene) map is in the centre and the physical cytological (pachytene) map on the right. Correlations are denoted by the dotted lines. Reproduced from Tanksley

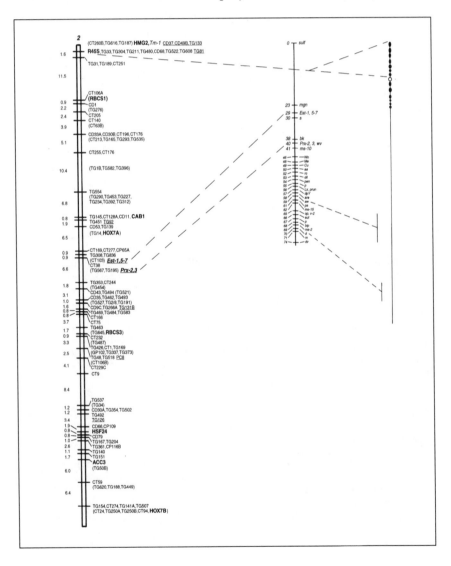

A rough guide for the human genome is 1 Mb to 1 cM. However, with chromosome 21, this becomes 200–300 kb to 1 cM and when compared with a plant, e.g. *Arabidopsis*, this alters again to 150 kb to 1 cM. Recombination also occurs non-randomly along a chromosome, in particular, regions near centromeres and telomeres exhibit suppression of recombination (in *Drosophila* this can be up to 40-fold suppression around centromeres) and hence appear to have fewer genes localized around them when the genetic map is produced. This suppression is thought to be caused either by the centromere exerting a direct effect on crossovers within the flanking regions or by the presence of heterochromatin. Conversely, there are also recombination hot spots, which distort the genetic map.

There are also sex differences in recombination frequencies in humans. Using chromosome 21 again as an example, the recombination length for males is 143 cM and for females 182 cM, with recombination in females being greater around the centromeres and reduced around the telomeres. It is not known whether this holds true for other mammals or why it takes place or the consequences.

Thus, all mapping data must be individually assessed taking into account not only the organism involved, but down to the level of chromosome regions and even to the molecular level. Complex computer programs have now been developed which allow for a certain measure of coordination between the two types of map.

7.5   Mapping strategies

The methods by which sequences can be allocated to a specific chromosomal region have been discussed, along with the tools for genetic mapping. But how do these relate to the process of defining genes which are responsible for disease or understanding mutational changes? It is very rare that a gene sequence can be isolated directly; the whole process is an interaction of genetic and molecular studies, involving a layered investigative approach. There are two main strategies for defining and cloning disease loci (Figure 7.15).

1. Functional cloning. This technique is used when the basic biochemical defect of a disease is known. It enables the purification of the protein involved and amino acid determination. This sequence can then be used to determine the RNA/DNA base composition and subsequently to clone the gene. For the vast majority of genes

Figure 7.15 Cloning strategies.

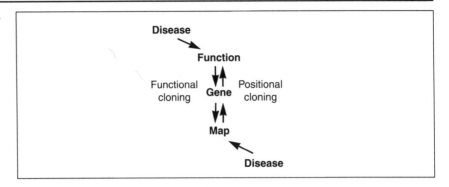

Box 7.6 Fine molecular mapping

Screening of affected families using pedigree analysis of linked DNA markers can narrow down a disease locus to a region encompassing 1–5 Mb of DNA. In molecular terms this is immense, and it is not practical to sequence the entire region. It has to be broken down into smaller pieces for analysis of specific regions in the search for candidate gene sequences.

Once linked markers have been defined for a disease, these can be used as probes on DNA libraries to isolate homologous clones. If all markers do not hybridize to the same library clone (very unlikely), a series of overlapping clones must be generated by chromosome walking. Chromosome walking utilizes the sequence of the end of one clone to probe by hybridization for the adjacent overlapping clone. A series of overlapping clones is called a contig (for contiguous stretch of DNA).

Generally contigs are constructed from either YAC or cosmid libraries. This is because these vectors can accommodate large pieces of foreign DNA (up to 2 Mb and 45 kb respectively) and are therefore the most efficient way of covering a large region (for more on YACs see Chapter 2). Because of their large size, the clones within these vectors often contain large stretches of repeated DNA. This can be a problem, particularly if present at the end of a clone, as they can effectively act as full stops in searching for adjacent clones. In a repeat sequence tract with potentially hundreds of similar sequences, often differing by only a few base pairs, it is often impossible to work out the orientation and direction of the walk through this region. Searching for candidate genes by the methods described later for cystic fibrosis (i.e. CpG islands and open reading frames) is generally carried out on cosmids and YACs. However, if potential candidate genes are found, characterization can only fully be carried out at the DNA base pair level, i.e. sequencing.

Most sequencing output can only produce 4–800 bp at a time, and so the larger pieces of DNA (in YACs and cosmids) are subcloned into plasmids. Sequencing is carried out in the plasmids, and gradually a series of overlapping sequenced clones are fitted together to produce the whole gene sequence. This may be of the order of several kilobases particularly if the genomic sequence is studied, as this will also contain all the introns and flanking and internal control sequences. It is a very labour-intensive and laborious process.

Not all fine analysis is carried out at such a detailed level. (Pulsed field gel electrophoresis (PFGE) is a specialized agarose gel technique that can isolate pieces of DNA up to 10 Mb (sufficient to isolate whole chromosomes from organisms with small genomes such as yeast, trypanosomes and *Plasmodium*). The fragments from PFGE are large enough such that several genes can be mapped at the molecular level by determining whether they co-migrate on the same piece of DNA when it is subjected to PFGE and Southern blotted. This technique is often necessary for some very important human genes such as dystrophin (mutations can produce Duchenne muscular dystrophy, a neuromuscular disorder), which is the largest human gene known at 2.4 Mb. Other examples include those for cystic fibrosis and Huntington's chorea, both in the 2 Mb size range.

this information does not exist, hence the second strategy is now generally often used.

2. Positional cloning. This is also known as reverse genetics. Pedigree analysis, i.e. screening of affected families, is carried out using multiple polymorphic DNA markers to establish co-segregation (linkage) between the disease and a region of DNA (this can be a very hit and miss approach as, if location is not even approximately known, the whole human genome has to be searched using markers and chromosomes gradually eliminated from the search). Once gross linkage has been established, the area affected can be narrowed down using more detailed markers and examined at the molecular level for candidate gene sequences (Box 7.6). The general approach is shown in Figure 7.16.

Each candidate gene has to be examined for mutations in the affected individuals and tissue-specific expression. Clear assignment of the

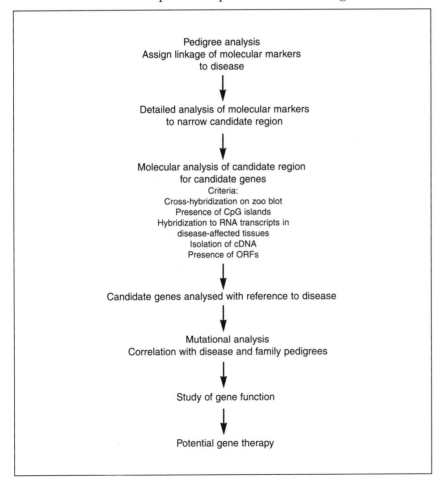

Figure 7.16 Flow chart of positional cloning strategies.

Box 7.7   Model organisms

The HGP includes in its aims the complete sequencing of several model organisms: yeast (*Saccharomyces cerevisiae*) (Chapter 2), *Caenorhabditis elegans* (a nematode), *Arabidopsis thaliana* (a plant), mouse, *Escherichia coli*, *Bacillus subtilis* and two mycobacteria.

Model organisms are used for many reasons.

1. They have a small genome size. This means that sequencing of the total number of genes which are responsible for the formation and maintenance of that organism is achievable. Their organization and regulation will be deciphered much faster than in humans (partly because they do not contain so much repetitive DNA).
2. The systems are much more easily manipulated for the production of mutants in mutation studies. This allows developmental and biochemical pathways to be elucidated and hence new genes can be isolated.
3. They can be used as experimental transgenic systems to define both the gene action itself and the surrounding control regions of a gene.
4. New genes found in these model organisms can be mapped and characterized and, if homology exists, their equivalent cloned in humans. The more evolutionarily distant the model organism, the greater the emphasis is placed on any preservation of sequence homology, i.e. if sequences are highly conserved over a wide evolutionary range this implies that their function is essential, in terms of both action and control.
5. Synteny (i.e. conservation of gene sequence) may exist over extensive regions, allowing the results of mapping from one model system to be directly applied to other organisms, including humans.

Each model organism has its particular uses, as follows.

- Yeast is used for analysing cell division and replication processes.
- *C. elegans* is used in the study of developmental biology in terms of cell positioning, function and pathways.
- *A. thaliana* is a higher plant with the smallest known genome (100 Mb). It has a rapid growth cycle, a minimum amount of repetitive DNA and prolific seed production. It is used for mapping purposes and isolating plant genes via mutagnesis, a process that would be much more difficult in the vast majority of plants, many of which contain up to 80% repetitive DNA.
- Mouse is a eukaryote and a mammal similar to man in terms of development, genome size and complexity. It also displays gross homology in the chromosomal arrangement of its genes. It is easily bred and hence a manipulatable system in which genes can be localized and studied.

Other model organisms exist and the list is being increased, rather than reduced, all the time. The best known species include rat (reasons similar to mouse), zebrafish (experimental embryology and genetic analysis), chicken and frog (embryology), *Drosophila* (developmental studies, regulatory sequences) and the Japanese puffer fish (*Fugu rubripes*) (evolution of genes and regulatory elements, plus intron size is much smaller in *Fugu*, by approximately one-seventh, thus making it easier to clone and analyse the function of huge genes such as the dystrophin and Huntington's, chorea genes via transgenics).

candidate gene as the causative agent in the disease is only really established with the demonstration of mutations (these may be in either the introns, exons or control sequences). Even then the work is still not finished and continues with extensive examination of the surrounding regions for promoter and control sequences and characterization of the gene product. This process is aided by comparative mapping studies using model organisms (Box 7.7).

The size of families in the genetic analysis limits the resolution of the pedigree analysis, usually to intervals of 1 cM or 1 Mb. In families with rare diseases this may be extended to 5 Mb. Other complications include the fact that many diseases, while having an inherited form, also occur sporadically; for example, two thirds of the cases of Duchenne muscular dystrophy (DMD) are inherited, whereas one-third are sporadic. Other common diseases for which all the genes have not yet been discovered include breast cancer. Although for some diseases major loci may exist, the effects may also be modified by other multiple loci that contribute small effects, e.g. diabetes, hypertension, coronary artery disease, etc.

An example of gene mapping where positional cloning has been successfully used, cystic fibrosis, is now examined in detail.

Cystic fibrosis (CF) is the most common severe autosomal recessive disorder in Caucasian populations. Five per cent of the population are carriers resulting in 1:2000 affected live births. It is thought to be due to abnormal regulation of chloride exchange channels in the apical membranes of airway and other epithelial cells. The result is inefficient clearage of mucus and micro-organisms from the airway, resulting in chronic lung disease. There are also associated gastrointestinal problems.

No chromosomal rearrangements or deletions could be detected in families affected with this disease, giving no initial clues as to the approximate location of the gene. Pedigree analysis was carried out on affected families using linkage analysis of polymorphic DNA markers. This enabled the CF locus to be localized to the band 7q31 (Figure 7.17).

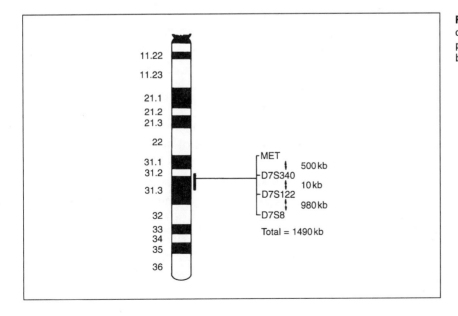

Figure 7.17 Long arm of chromosome 7, showing position of, and distances between, CF markers.

More detailed molecular analysis showed it to be associated with four markers: MET–D7S340–D7S122–D7S8, a total region covering 1490 kb. Genetic data indicated that D7S340 and D7S122 were in closest proximity to the gene. The basic idea was to clone the surrounding DNA and search for candidate gene sequences. A 280 kb region consisting of cosmid and phage contigs was constructed. The cosmids and phage were restriction enzyme mapped to orientate and align the different clones.

This region was then examined for the presence of candidate gene sequences. Four candidate regions were identified for further analysis.

Region 1 contained a potential open reading frame (ORF) and produced a 1.8 kb RNA transcript, but was dismissed on the grounds that it did not fit the genetic data.

Region 2 cross-hybridized strongly in a zoo blot. Extensive sequencing was carried out on this region, but database searching revealed close similarity to a region in the β-globin locus which was equivalent to a transcribed repetitive DNA family.

Region 3 was highly CpG rich. ORFs were detected, but no corresponding RNA transcripts could be detected or cDNA clones identified, indicating that it was perhaps a highly tissue or developmentally specific gene.

Region 4 showed strong cross-hybridization on a zoo blot, but very weak RNA transcription. The sequence was very GC rich and contained a short ORF. The GC region contained sites for *Bss*HII and *Sac*II, which are often found associated with undermethylated CpG islands (an indication that a transcribed sequence may have been found). Tests with methylation-sensitive restriction enzymes indicated that the region, as a whole, was undermethylated. Eventually a single cDNA clone was isolated and characterized. This proved to be the 5′ end of a gene. Overlapping cDNA clones produced a 6.5 kb transcript which was expressed in many tissues, but at particularly high levels in those tissues affected in CF patients. This candidates gene was then examined in affected families to determine the nature of any mutations which may cause the disease.

The amino acid sequence was determined in affected and unaffected individuals. It revealed a 3 bp deletion resulting in the loss of a phenylalanine residue at position 508. ΔF508 is involved in 68% of all CF mutations, but another 165 variants have since been discovered. This lack of a single causative mutation means that routine prenatal diagnosis may be limited, in the first instance, to those families with the ΔF508 mutation.

The isolation of the DNA sequence allows definition of a putative gene product (by comparison with similar sequences in the database). It has been called CFTR (cystic fibrosis transmembrane conductance regulator) and has been found to be membrane associated and involved in the transport of ions. Studies will continue to define the gene function accurately. So far, the mutations have been found to result in either mislocation of the protein or a reduced function. Having said that, gene therapy studies have already started on a limited scale.

Mapping at the chromosome or molecular level is only the first stage in the process of understanding gene action. Now we have the ability to refine the maps further by determining whether genes are chromatin loop or histone associated, how they interact with the chromosome scaffold and the consequences of this. This work, currently in its infancy, will provide much interesting further evidence on the nature of gene action and control.

Thus, mapping not only provides a list of genes, but is starting to tell us about the complex nature of the chromosome and how it acts.

8 Sex chromosome systems

Summary

Sex is a complex subject, with an almost infinite number of sex-determining mechanisms present in nature, the most familiar of which is the possession of sex chromosomes. In those organisms which have been shown to have such a specialized set of chromosomes, it is assumed that this is the primary sex-determining gene block, although the possibility that there may be modifiers present on other chromosomes cannot be discounted. The chromosomes carrying the sex-determining genes often differ in size, shape and staining properties. In vertebrates, sex chromosomes are common in mammals and birds, less so in reptiles and rare in amphibia and fish (Table 8.1).

Our knowledge of most sex systems is very limited. Cytogenetic evidence is often all that is available; experimental and gene mapping evidence is invariably lacking for the majority of species.

8.1 Heteromorphic sex chromosome systems

In mammals, the female usually has the homomorphic chromosomes (XX) and the male the heteromorphic set (XY). This situation is reversed in some other species, e.g. birds, in which the female is the heterogametic sex (ZW) and the male is homogametic (ZZ). The designation of Xs, Ys, Zs and Ws merely differentiates between male and female heterogamety (Figure 8.1). The XX–XY and ZZ–ZW systems are the

Table 8.1 Summary of sex chromosome systems

	Predominant sex chromosome system	Predominant heteromorphic sex	Occurence of sex chromosomes
Mammals	XX/XY	XY male	Universal
Fish	None	–	Sporadic
Amphibia	None	–	Sporadic
Reptiles			
Snakes	ZZ/ZW	ZW female	Almost universal
Lizards	None	–	Sporadic
Others	Temperature dependent	–	Practically zero
Birds	ZZ/ZW	ZW female	Universal
Insects	None	–	Sporadic
Plants	None	–	Rare

most common form of sex chromosomes but are not unique and many variants exist, as will be shown later.

In general, when only one pair of sex chromosomes is present, the heteromorphic chromosomes are highly differentiated and only share a small region of homology which pairs at meiosis (the pseudoautosomal region). Consequently, there is little exchange of genetic information between the two types of chromosome (Figure 8.2).

The heteromorphic chromosome (Y or W) is generally smaller, late replicating and partly or wholly heterochromatic. It contains few functional genes and shares very little homology with sequences on the homomorphic chromosome (X or Z) (apart from in the pairing

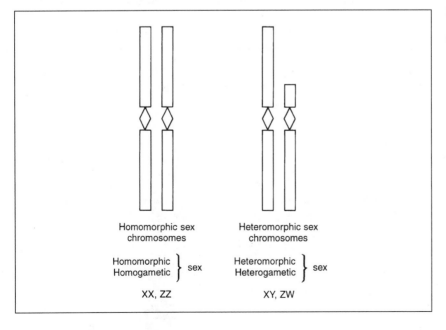

Figure 8.1 The main terms used in connection with sex chromosome systems.

Homomorphic sex chromosomes

Heteromorphic sex chromosomes

Homomorphic } sex
Homogametic

Heteromorphic } sex
Heterogametic

XX, ZZ

XY, ZW

Figure 8.2 Human X and Y chromosomes showing region of pairing at the very tip of the short arms (X, p22.3; Y, p11.3). Note that there is no banding homology. Reproduced with permission from O'Brien and Marshall-Graves (1991) *Cytogenet Cell Genet*, S. Karger AG, Basel.

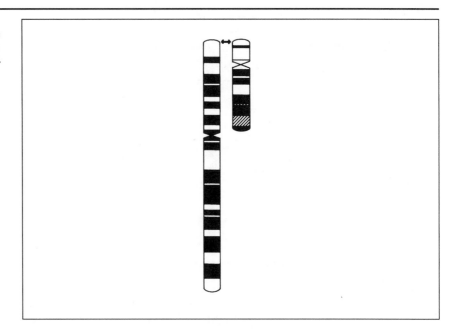

segment). The sex chromosomes, in general, contain a reduced quota of essential housekeeping genes compared with the rest of the complement. Hence, the organism can often tolerate far more numerical variation within this set than with the rest of its autosomes.

8.1.1   Evolution of heteromorphic sex chromosome systems

Chromosome evolution will be discussed as a whole in Chapter 9. However, sex chromosomes represent a special type of evolution as single chromosomes within the sex chromosome set have evolved separately, whereas in general terms evolution acts on both chromosomes of a pair as if a single unit. There is a huge variety in the structure and genetic differentiation of sex chromosome systems in many different genera; this diversity implies that sex chromosome heteromorphism must have evolved independently many times and therefore that simple evolutionary forces are involved.

The generally accepted view is that ancestral sex chromosomes were morphologically identical, with sex determined by a one-gene, two-allele system. Homozygosity determined the development of one sex and heterozygosity the other. This single gene system was not necessarily absolute, but part of a pathway that controlled sex determination in a coordinated manner.

Development of heteromorphism appears to have evolved via a series of gradual changes involving:

- partial or complete suppression of crossing over between the two homologous chromosomes in the heteromorphic sex;
- functional degeneration of the heteromorphic chromosome (Y or W).

In the generalized discussion, references to X and Y chromosomes also apply to Z and W chromosomes.

Crossover suppression

The reason for crossover suppression is not clear. It has been suggested that if sex-linked genes initially had opposite effects on the fitness of the two sexes, there would be a selective advantage in reducing recombination between the sex-determining locus and these genes in the advantaged sex.

A further reduction in genetic exchange over a wider area could be achieved if any further genetic variants (perhaps affecting physiological, morphological or behavioural differences) evolved which were advantageous in the heterogametic sex but detrimental in the homogametic sex. These variants would spread only if linked to the sex-determining region of the heteromorphic chromosome, thus providing a selective advantage in reducing recombination between the sex-determining locus and these genes in the heterogametic sex.

How might crossover suppression occur? There are several potential methods.

1. Genic. The selection for genes which reduce recombination may have been one of the initial factors, but would have been superseded by other, more wide-ranging, effects. It is difficult to see how this could effectively operate on one particular chromosome set and cause such large-scale isolation as seen in the X and Y.
2. Conformational changes include heterochromatization and alteration in the timing of replication compared with the autosomes (i.e. late replication). There are many known examples of sex chromosomes that are virtually indistinguishable except for a block of heterochromatin revealed by differential staining, e.g. some European newts of the genus *Triturus* and the fish *Poecilia sphenops* var. *melanistica* (Molly species). Some of the more primitive snakes lack morphologically distinct sex chromosomes, but one of a pair is seen to replicate late.

3. Structural changes such as peri- and paracentric inversions. The structural rearrangement itself, because of the effect of pairing in meiosis, partly acts as a crossover suppressor. This produces incomplete pairing at meiosis and, if recombination occurs in the inversion loop, aneuploid gametes are produced. To maintain fertility this change would have to be accompanied by selection for inactivation of this region (conformational changes). Inversions have been shown to be the only difference between the sex chromosomes of the gekko *Gehyra purpurascens*.

It has been suggested that crossover suppression is caused by a mixture of both conformational and structural changes. Which came first and is the most important is difficult to determine. Although examples can be found in many species in different stages of evolutionary development, the mechanism probably varies according to the species, as can be seen in the examples given, and represents many isolated incidents of independent evolution.

The effect of crossover suppression is that the two types of chromosome (X and Y) become partially isolated. As can be seen from the preponderance of small Ys (relative to the size of Xs) the path is then open for the degeneration of heteromorphic chromosome structure and function. This may occur by accumulation of deleterious mutations and DNA damage (such as deletions), ultimately, over time, leading to differentiation between the two types of chromosome.

Functional degeneration

Detrimental mutations do not tend to accumulate in the homogametic sex (XX) as recombination and repair can occur between the two Xs. This is not true of the non-homologous region of the Y chromosome. Although the rate of mutation in the X and Y chromosomes is the same, the rate of fixation is higher in the Y. The mechanism whereby this increased rate of fixation is achieved is based on 'Muller's ratchet' model (Box 8.1).

It is suggested that degeneration of the Y is also achieved via accumulation of heterochromatin. This process can take place once functional degeneration and crossover suppression have become established. Many Ys are heterochromatic, implying that there is some selectional advantage in this process. However, the potentially beneficial effects of heterochromatin in suppressing gene activity and reducing XY recombination cannot alone account for this.

Box 8.1   Muller's rachet

The ratchet (Figure 8.3) is a model for the progressive, one-way accumulation of mutations. It operates on a finite population with a steady mutation rate producing deleterious alleles. This population contains a variety of Y types with different numbers and types of mutations, but the effect of each mutation is small and only a slight selective advantage exists between chromosomes with different numbers of mutations. Mutant-free Y chromosomes can be lost from the population by random drift. They will not be replaced by new mutant-free Ys because they cannot be regenerated by recombination and the possibility of reverse mutation is infinitely small. The ratchet moves on a notch once these have been lost; the next generation of chromosomes with a low number of mutations can be lost in a similar way and the ratchet moves a further notch. This process will continue with the gradual accumulation of mutations in the Y. It has been suggested that this rate can be enhanced if, in the absence of recombination, selection of beneficial alleles which are linked to deleterious mutations occurs (genetic hitch-hiking).

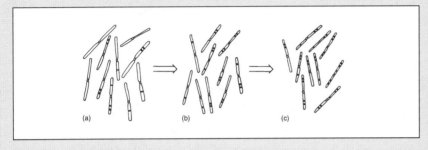

(a)   (b)   (c)

Figure 8.3   Muller's ratchet. Population A (demonstrated by chromosomes) has a mixed level of mutations (denoted by stripes) ranging from 0 to 3. The chromosomes with no mutations are gradually lost and cannot be regenerated, but more mutations will accumulate (population B with one-, two-, three- and four-mutation individuals). The individuals with one mutation are then lost from the population, replaced by individuals with five mutations (population C), thus moving the ratchet on another notch.

Another theory is that heterochromatin accumulation may be due to selfish or parasitic DNA. The Y chromosome accumulates DNA, whose only function is to propagate itself. This is not a serious disadvantage to the Y because any loss of genetic function is compensated for by the X. Loss of these additional sequences or correction is prevented by the lack of recombination.

Many Ys and Ws are smaller than their homomorphic partners, and it has been hypothesized that loss of material may be evolutionarily advantageous as this would reduce the energy costs associated with replication of this chromosome. Polymorphism of Y chromosome size, as for example in humans, is assumed to occur in species in which this degenerative process is not completed.

Accumulation of all these effects (genic, conformational and structural changes, mutations, etc.) leads to the structural and functional degeneration of the Y in the form of a cyclic process, as demonstrated in Figure 8.4.

Figure 8.4 Cyclic process of Y-chromosome degeneration.

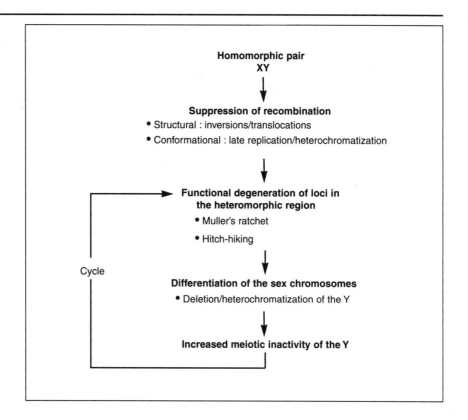

8.2 Mammalian sex chromosome systems

The vast majority of mammals have XX (female) and XY (male) sex chromosomes (Figure 8.5). Examples of multiple sex chromosome systems do occur, for example XY_1Y_2 in Indian muntjac, common shrew and long-nosed rat kangaroo. These are the exceptions rather than the rule, and this section will concentrate on the XX–XY system.

The lack of a great number of housekeeping genes on mammalian sex chromosomes means that there is a greater ability to tolerate aneuploidy compared with the autosomes. Most people have heard of the XXY/XXXY/X/XXX variants found in humans, however individuals with deviating sex chromosome constitutions have been found in a least 16 other organisms so far (Table 8.2).

The mammalian sex determination system is Y dominant. The primary gene switch for sex determination has been localized to the Y chromosome. It has been designated *SRY* for sex-determining region Y gene (referred to previously as TDF, testis-determining factor) and is

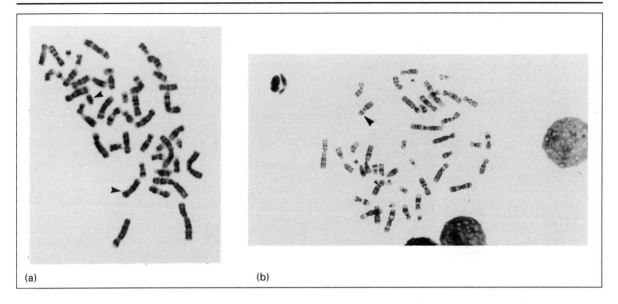

(a)   (b)

localized at position Yp11.3, close to the pseudoautosomal boundary (Box 8.2). Although this gene is the main sex-determining switch, the possibility that other genes scattered throughout the autosomes or X also play a role in sex determination has not been ruled out.

The pseudoautosomal region contains homologous sequences in both the X and Y chromosomes, and these pair at meiosis. The region directly proximal [containing genes for steroid sulphatase (STS), Xg blood group,

Figure 8.5 Banded cells from human (a) male and (b) female. Sex chromosomes are arrowed. Photographs courtesy of T. Spencer (a) and S.C. Rooney (b).

Table 8.2 Sex chromosome variants in mammals (from Fredga, 1988)

Sex chromosomes	Species found with variants
Female	
XO	Human, tammar wallaby, rhesus monkey, cat, horse, pig, cattle, sheep, South American field mouse, mole rat, black rat, mouse
XX	Normal
XXX	Human, horse, cattle
XXXX	Human
XXXXX	Human
Male	
XY	Normal
XYY	Human, brown rat, mouse, (common shrew, cat, horse, cattle)
XXY	Human, tammar wallaby, common shrew, dog, cat, pig, cattle, sheep, Chinese hamster, black rat, mouse
XXYY	Human
XXXY	Human
XXXXY	Human

Brackets denote mosaics.

Box 8.2 *SRY* gene

In human sex determination, female is the default pathway and, in general, possession of a Y chromosome alters this to male (the exceptions being where a small piece of Y chromosome has translocated to an X to produce XX males, for example). The Y chromosome codes for a gene which acts on the developing gonad to induce testis formation; the testis then triggers a cascade of various other gene actions to produce a male phenotype.

The position of the *SRY* gene was mapped by examining Y-chromosome fragments in XX males and determining the region of overlap. It is approximately 100 kb from the pseudoautosomal boundary. This close proximity to the pseudoautosomal region is thought to explain the frequency of X/Y translocations and sex-reversed XX males/XY females.

SRY has been isolated in several mammals, but only shows conservation of sequence over a 79 amino acid domain [high mobility group (HMG) domain]. This region has been found to bind DNA (consensus sequence A/TACAAT) and may therefore have a role in transcriptional activity of other genes (the targets of which are as yet unknown). The gene structure of the *SRY* region is shown in Figure 8.6.

The CGCCCGC sequence contained within the CpG island is a potential binding site for the EGR-1 and WT-1 family of transcription factors, some of which are functional in gonadal development. The consensus sequence for the HMG domain is found up to 10^5 times in humans, and therefore sequence specificity alone cannot define the biological action of this gene as it has been found to have a very limited expression pattern.

Any gene which is thought to play a major role in sex determination should be expressed in the embryo, before the differentiation of the genital ridge. In mice this process takes place at 12–12.5 dpc (days *post coitum*). *SRY* has been detected at least 2 days earlier than this and has been shown using *in situ* hybridization to be expressed only in the genital ridge region. *SRY* RNA in human males is limited to the adult testis. This precise temporal and tissue-specific expression of *SRY* implies that, although this may be the major gene switch for sex determination, there are other genes present which control its expression.

SRY obviously has an intimate relationship with sex determination in mammals. Its precise role, as yet, is undefined, as are the other genes it affects. It would appear that the higher order structure of the chromatin has a role to play, in tandem with sequence specificity, to define its action and control.

Figure 8.6 Basic molecular structure of *SRY* gene.

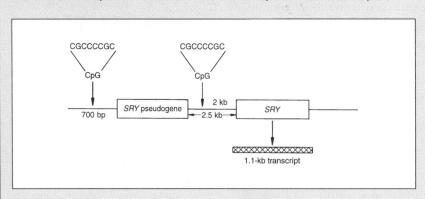

etc.) also shows extensive similarity between the two chromosomes. For example, the STS Y homologue contains 5 of the 10 exons of the active gene of the X and represents an unprocessed psuedogene. Other homologous sequences on the Y are also thought to be non- transcribed. Recombination between these homologous segments may be one of the main reasons behind some of the X–Y translocations and hence sex reversal found in humans (XX males and XY females).

Because each of the pair of sex chromosomes (X or Y) invariably contains different sets of active genes, most of which are on the homomorphic chromosome, the heteromorphic sex may well be at a disadvantage in having only one copy of these genes. Any alleles on the homomorphic chromosome, even if recessive, will be expressed. In humans several diseases have loci on the X chromosome, such as Duchenne muscular dystrophy, red–green colour blindness, fragile X-linked mental retardation and haemophilia A and B.

Overall estimated rates of incidence are:

Duchenne muscular dystrophy (DMD)	0.2/1000
Haemophilia type A	0.1/1000
Haemophilia type B	0.1/5000
Fragile X- linked mental retardation	0.1/1000

These are all recessive disorders, so (in theory) women need to inherit two alleles for the disease to be expressed (a comparatively rare phenomenon), whereas men will be affected with only one allele present. However, the situation is more complex. Where recessive disorders have been studied in detail, for example haemophilia, non-random X inactivation may give female carriers disease symptoms and so up until recently (with the advent of molecular analysis) these women would have been designated as homozygous suffering from a mild form of the disease.

8.3 X inactivation

X inactivation is a characteristic feature of the mammalian system. In 1961, Lyon and Russell independently postulated that during the early embryogenesis of female mammals, one of the two X chromosomes is randomly inactivated and becomes functionally heterochromatic. These ideas led to the development of the 'Lyon hypothesis'.

1. X-chromosome inactivation is random between maternal and paternally derived chromosomes.
2. It affects the whole chromosome.
3. It is irreversible in the cloned descendant of each cell, once it has occurred.

It is, in other words, the mechanism that mammals have evolved to cope with having a set of two Xs in the female (and hence two potential complete sets of X-linked expressed genes) but only one X in the male. X inactivation maintains the same dosage balance between the X and the autosomes in both males and females. The inactive X is observed as a heterochromatic body in interphase nuclei and is called a Barr body (Figure 8.7).

Figure 8.7 Barr body from human female. Courtesy of T. Spencer.

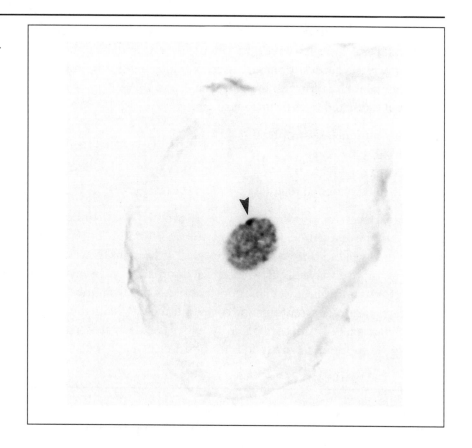

The Lyon hypothesis has since been found to be not entirely accurate. X inactivation is not completely non-random in all organisms. The classic example of X inactivation is in the marsupials, in which there is preferential inactivation of the paternal X chromosome and tissue-specific expression of some of the paternal genes. This is assumed to be a more primitive system than that in other mammals, as marsupials and man diverged 130 million years ago in evolution and marsupials are regarded as more primitive organisms. Current theory suggests that the mammalian random mosaic pattern of inactivation is more advantageous as it can prevent the full effects of harmful deleterious recessive genes.

In human females the inactivation phenomenon does not affect the whole X chromosome. A small portion at the tip of the short arm (Xpter–Xp22.3) is exempt. This includes the pseudoautosomal region and genes immediately proximal [STS, Xg blood group (Xg), Kallman's syndrome (KAL) and several loci coding for X-linked mental retardation (MRX) among others]. Several other genes in the regions Xp21.1–Xp22.2, Xp11.1–Xp11.3 and in the proximal long arm also escape inactivation. The reason is unknown, but the fact that several different regions are affected implies that factors other than chromosomal position determine the activity status of a gene.

What defines an active or inactive region is still largely unknown, as is the precise mechanism of inactivation itself. All models of X inactivation propose the existence of an X-inactivation centre (XIC) which is required for the *cis*-inactivation of the X chromosome. The molecular mechanism of inactivation is not well understood. However, it must involve a series of different stages.

1. Initial inactivation. XIC must be marked before inactivation as a means of distinguishing between the active X and the inactive X. Genomic imprinting and methylation have been implicated in this process.

2. Some method of counting. In individuals with multiple X chromosomes, all X chromosomes in excess of one are inactivated. It has been suggested that the active X is the default state with all other X chromosomes being responsive to the inactivation signal.

3. *Cis*-spreading of inactivation along the X chromosome. The inactivation signal leads to *cis*-limited transcriptional activity of genes located on either side of XIC. This is a difficult molecular model to deal with as inactivation has to be able to 'jump' over larger autosomal segments in X–autosome translocations, but not over to the other X chromosome (Figure 8.8).

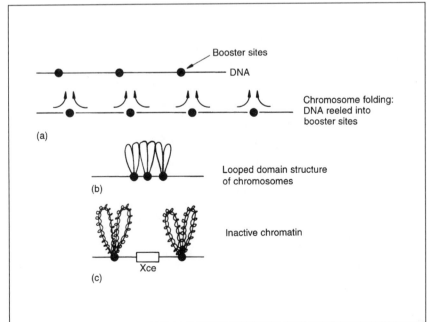

(a)

Booster sites

DNA

Chromosome folding: DNA reeled into booster sites

(b)

Looped domain structure of chromosomes

(c)

Xce

Inactive chromatin

Figure 8.8 Proposed X-inactivation mechanism. (a) Process of chromosome folding or 'reeling' of chromatin to form looped domains. (b) Normal active chromosome state. (c) A conformational change beginning at the Xce complex spreads over the chromosome at the base of the loops. The loops are more condensed on the inactivated chromosome. Reproduced from Riggs (1990), *Australian Journal of Zoology*, **37**, 432, Figure 7, with permission of CSIRO.

4. Maintenance of inactivation. Inactivation is stably maintained throughout most of the life of the female. DNA methylation of CpG islands and heterochromatization have been implicated in this process (as a means of switching off genes) but must have the ability to be reversed. Addition of methylase inhibitors inhibits X inactivation in cell culture, and reactivation of the X has been shown to be partly due to reduced methylation. Genome compartmentalization may also play a role. This is achieved by the late replication of the inactive X (there is growing evidence that early replication is required for transcription).

5. Reactivation of the inactive X in the oocyte and early embryogenesis. The mechanism of reactivation obviously depends on the mechanism of inactivation.

Cytogenetic and molecular studies on abnormal X chromosomes have identified a single region of the X containing the XIC (Xq11) and isolated a gene, *Xist*, which is strongly linked with XIC. This gene has been studied in both humans and mouse (Box 8.3).

The *cis*-limited nature of X inactivation is more readily explained by an RNA with a limited capacity for diffusion rather than by a protein. A protein would have to be exported from the nucleus, translated and then imported back into the nucleus: an unnecessarily complex process. This RNA could act in one of two ways.

8.3 *Xist* gene

The *Xist* gene is transcribed into a 15-kb RNA and is expressed only in the inactive X. *In situ* hybridization experiments with fluorochrome-labelled *Xist* RNA demonstrate its presence solely within the nucleus and concentrated on the X-inactivated Barr body.

This activity has been linked to the methylation status at the 5′ end of the gene. This region contains CpG islands; these are generally methylation free, but the *Xist* gene is the exception to the rule. Methylation at the CpG region extends through promoter sequences and 1.5 kb into the first exon of *Xist* in the active X chromosome (i.e. *Xist* expression is suppressed). *Xist* is expressed on the inactive X and shows no methylation of CpG sequences and promoters.

Studies in mice show that *Xist* expression precedes X inactivation by at least 1 day in mouse development. This implies that *Xist* expression may be involved in the actual initiation of X inactivation. This is backed up by experiments which show that methylation of *Xist* alleles precedes the onset of *Xist* expression.

Differential methylation and expression of the *Xist* gene follows a classic genomic imprinting pattern, similar to that of X inactivation.

The unique pattern of expression of *Xist* and its localization to·a key region for X inactivation in both human and mouse suggest that it is either involved in or directly influenced by the process of X inactivation.

Sequencing shows that *Xist* contains several tandem repeats; those at the 5′ end are evolutionarily conserved between humans and mouse. It does not contain any open reading frames of significant length and therefore does not appear to code for any proteins. This lack of a protein associated with the *Xist* transcript seems to indicate that if may function as a structural RNA within the nucleus.

1. It could interact in *cis* with the chromosome from which it is transcribed.
2. Transcription through *Xist* itself could induce conformation changes, allowing other factors to bind which facilitate the inactivation process. Inactivation could take the form of facultative heterochromatin formation or functional isolation within the nucleus, as part of the Barr body for example.

This theory of a structural role for *Xist* can be combined with a model for the actual mechanism of X inactivation and chromatin condensation. Because of the large distances involved (the X chromosome comprises approximately 150 Mb of DNA), it has been proposed that inactivation is not one single event but spreads by a series of booster sites located every 30–300 kb along the X by a reeling-type mechanism (Figure 8.8).

This reeling-type mechanism is based on observations of type 1 restriction enzymes (*Eco*B, *Eco*K). These enzymes bind to a specific site and then reel the DNA towards them, forming a loop. When two molecules of enzyme on the same strand meet and contact, the DNA loop is cut.

It is proposed that X inactivation works in a similar manner. The XIC forces a conformational change or releases an initiator, perhaps via *Xist* RNA, which starts the reeling process. This process has a limited distance effect and has to be enhanced at booster sites along the length of the X. When reeling has been completed in one region, it activates an adjacent region providing the appropriate booster sites exist (a domino effect). This causes the DNA to fold in a specific manner, which can then be maintained possibly by the scaffold proteins, ultimately resulting in the Barr body.

This is an attractive hypothesis since there is no free or long-distance diffusion of initiators required and it involves a *cis*-acting mechanism. The existence of booster sites could also explain why some regions are unaffected by inactivation. If these segments do not contain booster sites, the inactivation signal would pass through without effect.

The X-inactivation must then be stabilized and maintained. It is suggested that this is effected by a combination of heterochromatization, methylation and genome compartmentalization, as discussed earlier.

Some of these ideas are backed up by studies on marsupials, which have a more primitive form of X inactivation. All marsupial inactive Xs are late replicating and methylation is not used to a great extent in at least one marsupial species. This implies that perhaps genome compartmentalization came first, followed by methylation as a potential stabilizing factor. Ideas on the exact nature of X-chromosome inactivation are still at a rudimentary stage and much work needs to be carried out in the future to determine its true nature and action.

8.4   *Mammalian sex chromosome evolution*

It is thought that the mammalian X and Y diverged from a homomorphic pair in a common ancestor to mammals and reptiles (200 Myr ago). Conservation within the mammals for sex chromosomes is remarkable, whereas reptiles feature a variety of sex determination systems. This conservation is ascribed to the need to preserve the sex determination system and protect the dosage compensation mechanism of X-linked genes. Most research has centred around the mammalian X rather than the Y as this is the easiest of the two to evaluate, being, in general, larger, with more gene mapping information assigned to it.

Marsupial chromosomes can also be used to study sex chromosome evolution in mammals via comparative mapping. Mammals consist of three taxa: the eutherians (placental mammals), marsupials (including

Figure 8.9 Examples of marsupial Y chromosomes. (a) XY_1Y_2, swamp wallaby. (b) XY, kangaroo island wallaby. Note very small Y_1 and Y respectively. Courtesy of Professor D. Hayman.

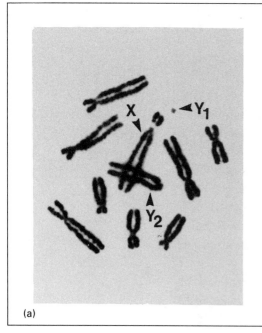
(a)

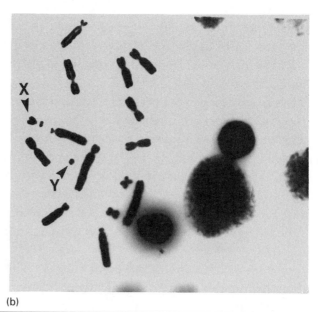
(b)

kangaroos and wallabies) and monotremes (consisting only of the platypus and two *Echidna* species).

Marsupial X chromosomes, in general, are smaller than the eutherian X and lack a pseudoautosomal region. The basic Y is very small and practically non-existent in some species (Figure 8.9). This seems to suggest that XY differentiation has gone to completion in many marsupials. X inactivation shows paternal imprinting and is incomplete in some tissues.

The monotreme situation appears more complex: the sex chromosomes pair over a larger area. It is proposed that the platypus has an XX/XY system and the echidnas a more complicated $X_1X_1X_2X_2$ female, X_1X_2Y male X/autosome translocation system (Figure 8.10). The situation is further confused by the appearance in both species of a multivalent chain involving the sex chromosomes in the first division of male meiosis and unpaired chromosomes in the mitotic karyotype. It has been demonstrated that Xp pairs with Yq at meiosis and that late

Figure 8.10 Female platypus karyotype. Courtesy of Professor J. Watson.

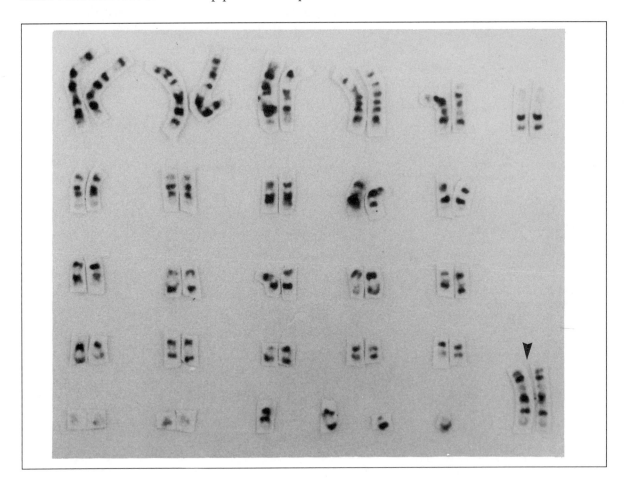

replication is confined to the unpaired region of the X in a tissue-specific manner.

It has been shown using comparative gene mapping studies that the original ancestral mammalian X consisted of the current human long arm of the X (Xq). The modern eutherian Xp genes were not found on the X chromosomes of any of the marsupial species studied, but localized to two autosomal clusters (Figure 8.11).

It is thought that the ancestral mammalian sex chromosome system consisted of a partially paired X and Y, with X inactivation restricted to a small differentiated segment (similar to the platypus system). Auto-somal segments were translocated to the pseudoautosomal region of the sex chromosomes in eutherians. The Y was gradually reduced in a stepwise fashion, accompanied by the spreading of the X-inactivation system over the newly unpaired regions of the X. The relatively recent addition of these autosomal segments could potentially explain why several regions of the human X are exempt from X inactivation, perhaps lacking the sequences responsive to the inactivation signal. Further work is continuing on this interesting group of animals which will further the knowledge of our own sex chromosome development and potential control mechanisms for X inactivation.

It is interesting to review the sex chromosome systems in other groups of organisms.

Figure 8.11 (a) Homology (shaded areas) between the X chromosomes of monotremes, marsupials and humans. (b) Assignment of Xp genes in humans, marsupials and monotremes showing origin of human Xp genes on autosomes of marsupials and monotremes. Data taken with permission from Marshall-Graves and Watson (1991) Mammalian Sex Chromosomes: Evolution of organization and function. *Chromosoma*, **101**, 65, Figure 2, copyright Springer-Verlag.

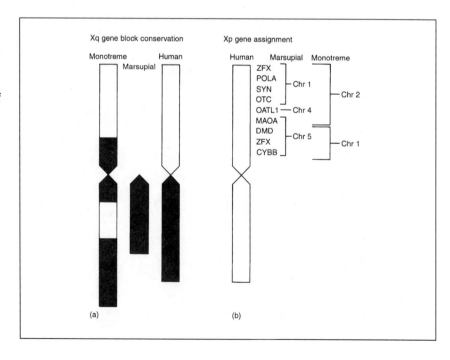

8.5 Other sex chromosome systems

8.5.1 Sex chromosomes in fish

Of all known fish species, only 2–3% have been cytologically examined; even fewer have identifiable sex chromosomes. Hence they are presumed to be at an early stage of differentiation along with amphibia and reptiles. One of the problems is that fish chromosomes are numerous and small, making it difficult to distinguish individual chromosomes. Sex determination in fish runs through the whole range of hermaphroditism, unisexuality (females producing only female offspring by gynogenesis or hybridogenesis), bisexuality, polygenic and, in one case, environmentally determined sex. The majority of fish are bisexual. With this wide range of sex determination mechanisms, it is not surprising that we find the sex chromosomes of fishes are relatively unspecialized and in a primitive state of development. The evolution of such systems is unknown, whether there is an increasing trend towards sex chromosome evolution or whether these are isolated adaptive cases, which evolved independently. There are almost certainly multiple origins of the development of sex chromosomes.

Genetic sex has been demonstrated in some species using sex reversal experiments and mapping of sex-linked genes (explained in more detail in amphibia).

Of the fish studied, there are examples of all types of sex chromosome systems, both more conventional (XX–XY, ZW–ZZ, XX–XO) and more complex multiple systems involving X–autosome and Y–autosome translocations ($X_1X_1X_2X_2$–$X_1X_1X_2$, ZW_1W_2–ZZ, $X_1X_1X_2X_2$–X_1X_2Y). It is difficult to generalize, as cases of proven sex chromosomes are so sporadic throughout the fish kingdom. Some families, such as anguilliformes (eels), contain only ZW–ZZ, whereas others, such as the Cyprinodontidae (subtropical/tropical carp-type fish with toothed jaws, e.g. guppy, sword tail) contain three different types (XX–XY, ZW–ZZ, $X_1X_1X_2X_2$–X_1X_2Y) but the number of samples is very limited (Table 8.3).

8.5.2 Sex chromosomes in amphibia

From the work carried out on amphibia, so far, the number of species with morphologically detectable chromosomes appears small when compared with other vertebrates. A broad-based survey of the sex-

	Scientific name	Common name	Sex chromosomes
Salmoniformes Bathylagidae (deep sea smelts)	Bathylagus wesethi		XX–XY
	B. stilbius	California smoothtongue	XX–XY
	B. ochotensis		XX–XY
	B. milleri	Big-scaled black smelt	XX–XY
Anguilliformes Anguillidae (eels)	Anguilla japonica	Japanese eel	ZW–ZZ
Congridae (conger eels)	Astroconger myriaster	White spotted conger	ZW–ZZ
Cyprinodontiformes Cyprinodontidae	Fundulus diaphanus	Killifish spp.	XX–XY
	F. parvipinnus	Killifish spp.	XX–XY
	Cyprinodon sp. (unnamed)		$X_1X_1X_2X_2$–X_1X_2Y
	Poecilia sphenops	Molly	ZW–ZZ
	Megupsilon aporus		$X_1X_1X_2X_2$–X_1X_2Y
Poeciliidae	Gambusia affinis	Mosquito fish spps	ZW–ZZ
	G. holbrook	Mosquito fish spps	ZW–ZZ
	Platypoecilus maculatus		ZW–ZZ

Table 8.3 Sample of sex chromosome systems found in fish. Note the paucity of samples from some species. Data from Ojima (1983) with permission

determining mechanisms and sex chromosomes of 61 amphibian species showed that 23 had heteromorphic sex chromosomes and 22 had homomorphic chromosomes when examined by plain staining. The heteromorphic sex was determined in all of these by either measurement of arm ratios, the amount of constitutive heterochromatin, loop patterns in lampbrush chromosomes or pairing in male meiosis. The remainder had no detectable sex chromosomes, but did have genetically differentiated sexes as determined by other methods (Box 8.4).

The main problem, particularly when studying sex-linked genes is that in most amphibia generation times are too long for crosses to be performed quickly. Also, of the mapping studies carried out to date, there appears to be no common ancestral conserved sex linkage group as can be demonstrated with mammals.

In those species possessing sex chromosomes, both types of XX–XY and ZZ–ZW are common and are present in both primitive and advanced families of frogs, toads, salamanders and newts. The stage of morphological differentiation is independent of evolutionary status. Within families both XX–XY and ZZ–ZW coexist, but within a genus they are the same type, indicating that the type of sex determination evolved after ancestral divergence (Table 8.4). There is no evidence of dosage compensation.

Some species are affected to a certain extent by temperature,

Box 8.4 Methods of genetic sex determination

HY antigen typing
This antigen is membrane associated and often found in a higher concentration in the heterogametic sex.

Segregation of sex-linked genes
This involves the study of expression and the patterns of inheritance of isozymes, some of which have been found to be sex linked, e.g. aconitase I, peptidase C and superoxide dismutase 1 in the frog genus *Rana*.

Sex reversal and breeding experiments (Figure 8.12)
Sex can be reversed in several organisms by hormone treatment (chemicals, e.g. oestradiol, methyl-testosterone, etc.) or implantation of the primordium of a testis at the site of an ovary in the embryo. These sex-reversed individuals are then mated with the same genetic sex and the sex ratio of offspring determined. Examples include the salamander family: *Ambystoma mexicanum* and *A. tigrinum*.

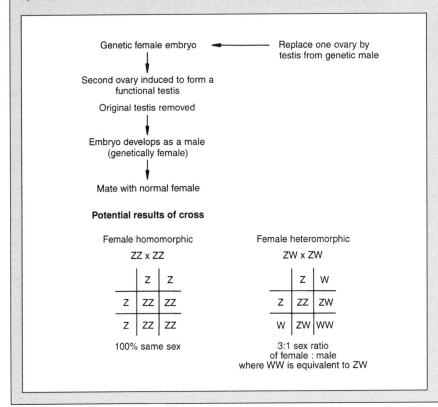

Figure 8.12 Demonstration of genetic sex determination by sex reversal experiments.

hormones, etc., and so potentially have some component of environmentally determined sex (described in more detail in reptiles).

In a more detailed survey of 833 of the known 3521 species of frogs and toads, only 11 possessed sex chromosomes; of these seven were XX–XY, three were ZW–ZZ and one, *Leiopelma hochstetteri*, possessed a very unusual WO–OO system in which the sex chromosome is a univalent supernumerary chromosome with a distinctive C-banding pattern. It is larger than the other supernumeraries and has a distinctive lampbrush structure. The heterochromatin distribution varies between

Table 8.4 Amphibian sex chromosomes. Data taken with permission from Schmid *et al.* (1991)

		Scientific name	*Common name*	*Sex chromosomes*		
Pipidae	(clawed toads)	*Xenopus laevis*	African clawed toad	ZZ–ZW		
Discoglossidae	(midwife/fire-bellied toad)	*Bombina orientalis*	Oriental fire-bellied toad	XX–XY		
Ranidae	(frogs)	*Rana brevipoda*		XX–XY		
		R. clamitans	Green frog	XX–XY		
		R. esculenta	Edible frog	XX–XY	XX–XY	X = Y,r
		R. japonica	Japanese agile frog	XX–XY		
		R. nigromaculata		XX–XY		
		R. pipiens	Leopard frog	XX–XY		
		R. ridibunda	Marsh frog	XX–XY		
		R. rugosa	Japanese wart frog	XX–XY		
		R. temporaria	European common frog	XX–XY		
		Pyxicephalus adspersus	South African bullfrog	ZZ–ZW	ZZ–ZW	Z > W
		Tomoptema delalandii	Small African bullfrog		ZZ–ZW	Z = W,i,h
Hylidae	(tree frogs)	*Hyla arborea japonica*		XX–XY		
		Gastrotheca riobambae	Marsupial frog spps		XX–XY	X < Y
		G. pseustes	Marsupial frog spps		XX–XY	X = Y,h
Bufonidae	(toads)	*Bufo bufo*	European common toad	ZZ–ZW		
Leptodactylidae		*Eupsophus migueli*		XX–XY		X = Y,i
Proteidae	(mud puppies/olms/ water dogs)	*Necturus alabamensis*	Alabama waterdog		XX–XY	X > Y
		N. beyeri	Gulf coast waterdog		XX–XY	X > Y
		N. lewisi	Neuse river waterdog		XX–XY	X > Y
		N. maculosus	Mud puppy		XX–XY	X > Y
		N. punctatus	Dwarf waterdog		XX–XY	X > Y
Ambystomatidae	(mole salamanders)	*Ambystoma mexicanum*	Axoloti	ZZ–ZW		
		A. tigrinum	Tiger salamander	ZZ–ZW		
Salamandridae	(newts)	*Pleurodeles walti*	Sharp-ribbed next	ZZ–ZW	ZZ–ZW	Z = W,lbc
		P. poireti	Poirets ribbed newt		ZZ–ZW	Z = W,lbc
		Triturus alpestris	Alpine newt	XX–XY	XX–XY	X = Y,h
		T. cristatus	Crested newt		XX–XY	X = Y,h
		T. helveticus	Palmate newt		XX–XY	X = Y,m
		T. italicus	Italian newt		XX–XY	X = Y,h
		T. marmoratus	Marbled newt		XX–XY	X = Y,h
		T. vulgaris	Smooth newt	XX–XY	XX–XY	X = Y,h
Plethodontidae	(lungless salamanders)	*Aneides ferrus*	Clouded salamander		ZZ–ZW	Z = W,i
		Chiropterotriton abscondens	–		ZZ–ZW	Z = W,i
		C. bromeliacea	–		XX–XY	X > Y
		C. cuchumatanos	–		XX–XY	X > Y
		C. rabbi	–		XX–XY	X > Y
		Bolitoglossa subpalmata	–		XX–XY	X > Y
		Hydromantes ambrosii	Cave salamander spp.		XX–XY	X > Y,i,h
		H. flavus	Cave salamander spp.		XX–XY	X > Y,i,h
		H. imperialis	Cave salamander spp.		XX–XY	X > Y,i,h
		H. italicus	Eastern cave salamander		XX–XY	X > Y,i,h
		H. sp. nova	Cave salamander spp.		XX–XY	X > Y,i,h
		Oedipina bonitaensis	Tropical worm salamander spp.		XX–XY	X > Y
		O. poelzi	Tropical worm salamander spp.		XX–XY	X > Y
		O. syndactyla	Tropical worm salamander spp.		XX–XY	X > Y
		O. uniformis	Tropical worm salamander spp.		XX–XY	X > Y
		Thorius pennatulus	Mexican dwarf salamander spp.		XX–XY	X> Y
		T. subitus	Mexican dwarf salamander spp.		XX–XY	X > Y

Abbreviations:
i, differ on centromeric index;
h, differ on amount of constitutive heterochromatin;
r, replication banding pattern;
lbc, loop patterns in lampbrush chromosomes;
m, pairing in male meiosis.

populations, and there are also varied structural forms, indicating a more rapid evolution than the autosomes.

It is thought that, in the amphibia, primary morphological differentia- tion of the sex chromosomes was initiated by accumulation of repetitive sequences and heterochromatization, rather than by structural changes, as implicated in mammalian sex chromosome evolution.

8.5.3   Sex chromosome systems in reptiles

Early reptiles emerged from amphibian stock. Sex chromosomes have been found in many snakes and lizards. These sex chromosome systems include both male and female heterogamety with different degrees of sex chromosome heteromorphism. If distinct sex chromosomes are not found, genotypic sex is difficult to determine in reptiles. Two of the experimental techniques used with other organisms, breeding experi- mentally sex-reversed individuals and detection of sex-linked markers, are not practical with reptiles. Data suggest that the general evolutionary trend is from homomorphic to heteromorphic sex chromosomes as homomorphic chromosomes seem to be the ancestral condition. Rep- tiles can be divided (for the purpose of sex chromosome studies) into three groups: snakes, lizards and other groups.

Snakes

Phylogenetically, snakes are grouped according to their skeletal charac- ters. Boas and pythons (boids) possess the primitive ancestral type, colubrids (e.g. rat, tree and water snakes) are intermediate and vipers possess the most sophisticated skeleton type. The evolution of sex chromosomes also follows this general pattern: from homomorphic chromosomes in the boids, through the colubrids which mainly show female heterogamety (the Z and W are often the same size, but differ in the position of the centromere) to the vipers, which show female heterogamety with morphologically distinct sex chromosomes (Figure 8.13). The sex chromosomes are generally the fourth largest pair of chromosomes. The Z is usually the same size and shape in most snakes, but the W varies both within and between species.

Lizards

Sex chromosomes have been identified in seven lizard families, but these examples occur sporadically and the heterogametic sex varies between

Figure 8.13 Sex chromosomes (fourth chromosome pair) of the three groups of snake.

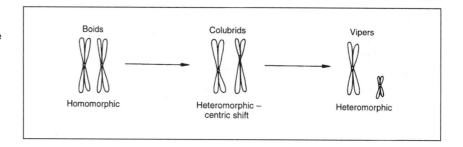

them. The sex chromosome pair is also not very differentiated, indicating their recent evolution (Table 8.5).

This variation in sex chromosome systems indicates multiple origins, at least three by recent studies, rather than the single ancestral type of the snakes. One problem with lizards is that the sex chromosomes are often microchromosomes and therefore can be difficult to detect. The results in Table 8.5 therefore represent minimum estimates for the incidence of sex chromosome heteromorphism.

Other groups

Looking at the rest of the reptile families, only one other case of sex chromosome heteromorphism has been identified, in the mud turtle family (the only difference between the pair of chromosomes was a terminal heterochromatic knob). It is becoming increasingly apparent that in many of the other members of this order – crocodiles, alligators, the iguana and Gecko families of lizard and in several families of turtles – these species do not have identifiable sex chromosomes and sex is primarily determined by the temperature at which the eggs incubate. So

Table 8.5 Sex chromosome systems in lizards. Data taken with permission from Bull (1980)

Taxon	Common name	Number karyotyped	Number with sex chromosomes	Heterogametic sex
Gekkonidae	Geckos	54	2	Female
Pygopodidae	Snake lizards	6	5	Male
Iguanidae	Iguanas	145	45	Male
Agamidae	Chisel teeth lizards	19	0	–
Chamaeleontidae	Chameleons	36	0	–
Xantusiidae	Night lizards	10	0	–
Teiidae	Whiptails and racerunners	46	1	Male
Lacertidae	Wall and sand lizards	33	4	Female
Scincidae	Skinks	35	3	Male
Anguidae	Anguids	12	0	–
Varanidae	Monitor lizards	18	4	Female
Amphisbaenia	Worm/ringed lizards	28	0	–

Box 8.5   Temperature-dependent sex (TDS)

This term is interchangeable with environmentally determined sex (EDS).

In general, low temperatures (22–27°C) produce one sex and temperatures above 30°C produce the other sex, with an intermediate 1–2°C range producing both; intersexes are rare. For example, American alligator (*Alligator mississippiensis*) eggs incubated at or below 30°C are female and at above 33°C are male. Male-producing temperatures in lizards are the reverse to those in turtles. So there is no general rule as to which temperature range produces which sex.

Single-sex nests are rarely produced, as there is usually a temperature gradient within the nest. Unlike chromosome sex determination, which produce a 50:50 ratio of males to females, TDS often produces strongly biased sex ratios, for example in most crocodiles this may be in the order of eight females to each male. The critical period for sex determination is within the second third of the incubation period.

There are some interesting arguments as to the adaptive significance (and disadvantages) and evolutionary role of TDS. It has been found that temperature controls more than sex; it also strongly influences how the young develop in the egg, and after hatching a nest with a range of temperatures produces a population adapted to a range of environments, so that some individuals of both sexes will do well under most environmental conditions. For example, in alligators higher incubation temperatures produce the largest males, and it is they who control harems of females. The larger the male, the bigger the harem, hence the larger alligators can mate for longer and pass on more of their genes. Most sexually mature females mate, so size is largely irrelevant. Although higher temperatures produce larger males, these require higher body temperatures and are more sensitive to stresses in temperature and food availability, so they will not reach their full potential under less than perfect conditions. Males produced at lower temperatures are smaller with a lower metabolic rate. They may be less successful in mating when conditions are ideal, but will thrive under poorer environments. These temperature-dependent changes can be viewed as an adaptive mechanism to different and changing environments. Such adaption would take longer with classical selection mechanisms.

The discovery of TDS has also provoked debate as to whether dinosaurs died out because they lacked this adaptability mechanism. Descendants of other amphibia and reptiles around at the same time still exist, but this may be because of their enhanced adaptability as a result of TDS. Dinosaurs with genetically determined sex could not adapt fast enough under classic selection procedures.

---

in this order, a different type of sex determination is in operation: temperature-dependent sex (Box 8.5).

Chromosomal and temperature-dependent sex exist in closely related species in both turtles and lizards, which suggests that it is evolutionarily possible to shift from one type to the other and that there may well be genetic components common to both. An interesting point is that with temperature-dependent sex a 1:1 sex ratio can be produced at intermediate temperatures, begging the question 'how can both sexes develop at the same temperature if no means of genetic differentiation exists?'. In three species (a snake, turtle, lizard) the sex ratio is usually 1:1 regardless of temperature, so potentially genetically determined sex may exist in these species. One theory is that genetic mechanisms exist, but that these are generally masked by more dominant temperature-dependent mechanisms.

8.5.4　Sex chromosome systems in birds

It is widely believed that birds have arisen from a reptilian lineage. The karyotypic arrangement of birds and their sex chromosome constitution is shared with some reptiles, notably the vipers, which also have sex chromosome differentiation. The avian karyotype has a characteristically large diploid number, the average being 80. There are also two chromosome size groups, the larger macrochromosomes being usually fewer in number than the smaller microchromosomes (a feature also shared with some reptilian groups).

Many species of bird do not exhibit sexual dimorphism (phenotypically differentiated sexes) either as young or adults, however most birds studied so far have a ZZ–ZW sex chromosome system with female heterogamety (Figure 8.14).

In most cases the Z and W are highly differentiated with appreciable differences in size. An exhaustive review of bird cytogenetics indicated strong evolutionary conservation of morphology between the sex chromosomes of birds, with the Z invariably being the fourth largest pair of macrochromosomes. There are, however, groups which show morphological variation of the Z chromosome even between closely related species. In the majority of species the W is a microchromosome and highly heterochromatic.

The exceptions to this general rule are the ratites (or flightless birds,

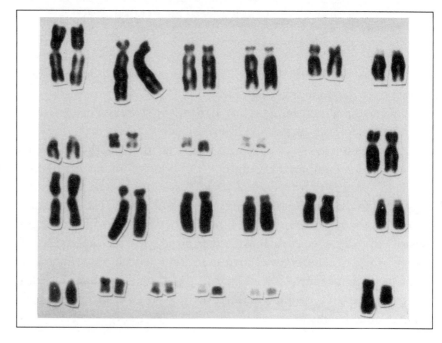

Figure 8.14　Karyotypes of male and female goldies lorikeets (*Trichoglossus goldiei*). Material prepared from cultures of feather follicle pulp. Only the macrochromosomes are shown (autosomes 1–7) with the three largest pairs of microchromosomes. Sex chromosomes (ZZ in male and ZW in female) positioned at the end of each karyotype. 2*n* = 68 in this species. Courtesy of S. Joshua.

e.g. emu, ostrich, rea). Plain staining of the chromosomes of these species shows no heteromorphic pair, however banding techniques have succeeded in identifying the Z and W chromosomes. Both chromosomes are euchromatic and exhibit similar banding patterns, indicating that they are in the early stages of differentiation. It is proposed that the evolution of the avian W chromosome has evolved via structural changes and loss of genetic material followed by heterochromatization, rather than the reverse process indicated in reptiles.

There is no evidence from banding studies and sex-linked loci that the avian Z is homologous to either the mammalian X or reptilian Z chromosome. Also, the Z chromosome does not appear to exhibit dosage compensation. Studies on the aconitase isozyme of four distantly related birds showed that males had twice the activity of the females. This finding was substantiated several years later when DNA replication studies demonstrated no asynchrony of replication between the two Z chromosomes in the males of three different species.

Although this implies no dosage compensation, it does not rule out the existence of small asynchronously replicating regions along both Z chromosomes. The W chromosome, as expected, due to its heterochromatic nature, was late replicating in all cases studied.

The lack of morphological sexual differentiation in many birds, for example Parrots (Psittaciformes), can obviously cause problems in captive breeding, so cytogenetic studies are a relatively straightforward solution to determining the sexes. Cytogenetic analysis using a non-invasive procedure to obtain mitotically active cells from growing feathers, has proven to be more reliable and carry fewer risks than examining gonads using endoscopy. The general difference in size between the sex chromosomes has also led to the proposal of using flow cytometry as a means of sex identification.

8.5.5 Sex chromosome systems in insects

As with fish, it is very difficult to generalize about a group as diverse as the Insecta.

Everyone at some point comes across that model 'animal' *Drosophila* and is aware that it has an XX–XO sex chromosome system, with sex determination based on X–autosome ratio. An X–autosome ratio of 1X:2A produces males and 2X:2A produces females. Any intermediate ratios produced result in intersexes. *Drosophila* also has a form of dosage compensation in which the male single X produces twice the transcription level of each female X. This sex chromosome system is also the most common found in the Orthoptera (e.g. grasshoppers), although occasional X–autosome fusions have occurred to produce neo-XY sex-determining mechanisms.

The other classic example is the Lepidoptera (e.g. butterflies), which have the ZZ–ZW system similar to birds.

However, other groups such as the Heteroptera (e.g. bed bugs and water bugs) contain many types of sex chromosome systems. XY systems are the most common, followed by XO and then multiple chromosome systems derived by fragmentation and translocation of the sex chromosomes and autosomes. A total of 1145 species have been examined, constituting 42 families. Of these, 846 are XY, 95 XnY (where n can be up to 4), 173 XO and 31 XnO (where n=2).

Compare this with examples from the Diptera, which often have a sex-determining linkage group that is not fixed on a particular autosome either within species or between species. The most studied dipteran example is the house fly, which has the XX–XY system in northern latitudes and autosomal sex determination in southern Italy. The zone between these locations has mixed populations, with the sex determiners found on different autosomes according to the individual populations. Within the Diptera different chromosome pairs serve as the sex-determining pair in different species or even within a species. Because of this wide variation, it is suggested that transposition may have played (and still is playing) a role.

On a final note, some insects to a large extent give up on sex and opt for parthenogenetic reproduction (the production of an embryo from an unfertilized female gamete). The main example is the Hymenoptera (e.g. honey bees and ants). These species have no discernible sex chromosomes and their sex determination can be broadly divided into two types: arrhenotoky, in which the unfertilized eggs develop parthenogenetically into males and fertilized eggs produce females; or thelytoky, in which unfertilized eggs develop into females.

8.5.6 Sex chromosome systems in plants

The vast majority of plants are hermaphrodite, producing both pollen and ova on the same plant. The association of dioecy with chromosome heteromorphism is extremely rare when compared with the situation in mammals.

A survey of over 120 000 flowering plants estimated that:

70% are hermaphrodite

5% are monoecious

5% are dioecious

2% are andromonoecious (plant with hermaphrodite and male flowers)

3% are gynomonoecious (plant with hermaphrodite and female flowers) and

15% are mixtures of the above.

Given that most plants are hermaphrodite, there are various mechanisms to ensure sexual outcrossing.

1. The development of the sexual organs can occur at separate times on the same plant (asynchronous hermaphroditism).
2. Fertilization be restricted by complex incompatibility interactions and male sterility.
3. Distinct organs may exist on the same plant (monoeocy) or on separate plants (dioecy). Examples of this latter category include hop, spinach, asparagus and willow.

Heteromorphic sex chromosomes have been claimed for many species, but in a recent survey on flowering plants their presence was substantiated in only five families, comprising nine species. As a generalization on the limited number of examples, it can be concluded that: all the sex chromosomes are the largest in the complement; the Ys or the sum of the Ys are larger than the X; and where comparisons are possible the Xs and Ys are the largest chromosomes in the genus. This is in complete contrast to mammalian systems, where the Y is much reduced (Table 8.6).

The study of sex chromosomes in plants has been of an extremely limited nature. Westergaard published a review in 1958, and his results

Table 8.6 Sex chromosome systems in plants. Data taken with permission from Parker (1990)

| Species | Chromosomes | | Sex determination |
	Auto	Sex	
Humulus lupulus (hop)	18	XX–XY	X:A ratio
	16	$X_1X_1X_2X_2–X_1X_2Y_1Y_2$	X:A ratio
Humulus japonicus (Japanese hop)	14	$XX–XY_1Y_2$	X:A ratio
Cannabis sativa (hemp)	18	XX–XY?	X:A ratio
Silene latifolia and *dioica* (red and white campion)	22	XX–XY	Dominant
Coccinia indica (Indian gourd spp.)	22	XX–XY	Dominant
Rumex hastatulus (sorrel spp.)	8	XX–XY	X + Y
	6	$XX–XY_1Y_2$	X + Y
Rumex acetosa (sorrel spp.)	12	$XX–XY_1Y_2$	X:A ratio
Viscum fischeri (mistletoe spp.)	14	4II/chain IX	?

Figure 8.15 Y chromosome of *Melandrium*. Segment I contains female suppressor genes. Segment II contains genes which initiate anther development. Segment III contains genes controlling the last stages of anther development. Data taken from Westergaard (1958), *Advances in Genetics*, **9**, 253.

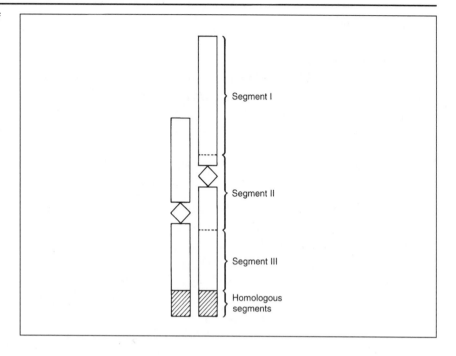

on the active Y system of *Melandrium* have only now been verified and the work expanded upon in more molecular terms. Sex expression in *Melandrium* depends exclusively on the strength of male determiners present in the Y chromosome. In fact, the Y chromosome can be segmented according to its function in sex determination (Figure 8.15).

In *Rumex acetosa*, probably the most studied species, it has been shown that sex is determined by X–autosome ratio (Figure 8.16).

The male sex chromosomes (XYY) form a trivalent at meiotic cell division and always segregate regularly with the X going to one pole and the two Ys to the other. Experiments have shown that only a quarter of the X is required to give a fully functional female and so far, despite its large size, only the Y has been shown to be essential for pollen fertility.

Sex determination in *Rumex* can be affected by the presence of extra autosomes, for example in triploids. Also, the Y dominance in *Melandrium* can be overcome by infecting female plants with the smut fungus *Ustilago violacea*, which somehow induces the female to develop anthers. Presumably the fungus is producing a sex hormone normally emitted by the Y chromosome, the nature of which, as yet, is unknown.

This is a phenomenon of all the plants listed: although there is a dominant sex determination mechanism in the diploid, this is a rather plastic phenomenon that can easily be altered by changing the

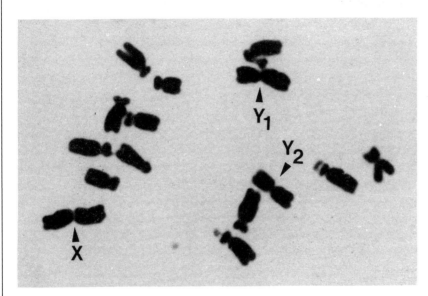

(a)

Figure 8.16 *Rumex acetosa* (sorrel) sex chromosome system. (a) Male chromosomes, X and Ys arrowed. (b) Female chromosomes, Xs arrowed. (c) Male meiosis with sex trivalent. Reproduced with permission from Parker and Clark (1991), *Plant Science*, **80**, 81–83, Elsevier Scientific Publishers Ireland Ltd.

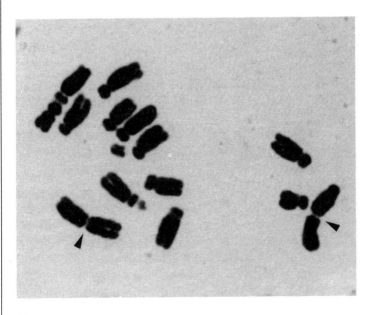

(b)

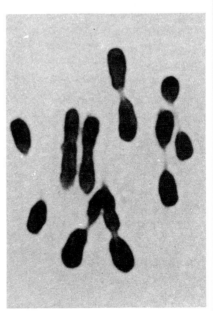

(c)

chromosome constitution of the autosomes, the background genotype or the addition of hormones.

Most current research concerning sex in plants focuses on the molecular events surrounding sex expression rather than the molecular genetics of sex chromosome systems. This is probably because this latter are very much in the minority and all the major crop plants are hermaphrodite. However, it has been suggested that investigating the sex determination system in plants with manipulatable sex chromosomes will prove a better means of investigating sex in plants in general.

Evolution and speciation

9

Summary

Two aspects of the process of speciation are of interest in the context of cytogenetics. The first of these is changes in ploidy i.e. changes in the number of the chromosomes, which themselves remain unaltered. Changes in ploidy can have both genetic and phenotypic effects, such as fertility changes, and can be used to great effect in plant breeding to produce new cultivars. Additional sets of chromosomes can be from the same individual (autopolyploids) or from an organism of genetically distinct origin (allopolyploids). In the case of humans, changes in ploidy can have very severe consequences, as can the second process of interest: changes in karyotype. Karyotype changes can be thought of as being due to changes in either DNA content or chromosome structure as well as changes in chromosome numbers. In this context it is possible to see speciation and karyotype changes that are linked, as in the marsupials, or not linked, as in the hominids.

Evolution and speciation are closely related to observable changes in an organism's chromosomes. It should, however, be clearly borne in mind that karyotype changes are rarely enough for speciation to occur on their own. It is, after all, the phenotype expression of the genome which determines the position of the fine line, sometimes indefinable, between variation and speciation. This chapter will discuss these general principles of evolutionary biology.

Changes in ploidy are a significant aspect of speciation, although whether or not they cause speciation is sometimes obscure. Chromosomes can introduce an effective barrier to interspecific crosses by (i) producing infertile hybrids and (ii) causing problems during cell division.

Construction of phylogenies based upon chromosome shape and size is difficult. Often such taxonomic relationships have been shown to be in very close agreement with phylogenies constructed from both morphological and molecular data.

9.1   Hybrids and cultivars

9.1.1   Changes in ploidy

The process of speciation is undoubtedly complicated, but it is possible to draw general inferences from observation of some organisms and their chromosomes. It is also possible to hazard a guess as to the relationships between different species based on cytogenetic data. Box 9.1 considers species numbers and diversity.

Polyploidy has been of considerable importance in speciation. This is particularly true in plants, for which estimates of polyploidy range from 30% to 80% in angiosperms. Before looking at specific examples it is worth thinking about the fundamental biological consequences of ploidy. As animals demonstrate a far greater level of structural and metabolic complexity than plants, it is easy to understand why ploidy changes are more important in speciation in plants than in animal speciation. Basic changes associated with an increased chromosome complement are an increase in cell size and an increase in the copy number of genes. The increase in chromosome number is directly correlated with cell size; consequently, it has proved possible to use fossil evidence to calculate the basal chromosome number of angiosperms. By measuring stomatal guard cell size and relating this to chromosome number in living forms, the guard cell size of fossilized plants can be used to determine ancestral chromosome number. Using

---

Box 9.1 Species numbers

Consider this: so far as it is possible to tell, the universe is not infinite, consequently it does not have an infinite mass. In the same way as the universe does not have an infinite mass there are not an infinite number of permutations based upon the sequence of DNA. Do not confuse the term 'infinite' with 'a very large number.'

So why are there so many species and how do we know how many species there are? This is a question which is fundamental to biology and evolution.

Although there is much debate as to the exact number of species that are extant upon the planet, it is possible to develop ideas about the biodiversity of the world. It may, erroneously, be suggested that it is not important to know how many species there are. It is, however, axiomatic that if you do not know how many species there are you cannot know how many have become extinct. It is almost as fundamental a value to biology as Planck's constant is to physics.

It is possible to estimate the number of species becoming extinct, but not the number which are arising. The level of diversity that is apparent from the number of identified species can be misleading. The most fashionable groups will be the most studied and have the most named species, and the groups with the largest individuals will have the fewest undiscovered species. When J.B.S. Haldane was asked what a lifetime in biology had taught him about God he replied 'He has an inordinate fondness for beetles'. This is a reflection of the number of described species. At the moment the number of described species of insect outnumbers all other species put together. Although the class Insecta is very successful there is no fundamental reason to believe that the Coleoptera is intrinsically more successful than any other insect order; it is just that more have been identified. With this uncertainty it seems imperative that a biodiversity audit of the planet is undertaken.

However, several groups are sufficiently well known to be almost completely counted, so although the numbers may vary slightly with the occasional discovery of new species we know there are 9881 bird species and 4327 mammal species. This is out of an estimated 1 770 000 described species, which is much less than the actual total number of species. Estimates of total numbers vary from 5 to 30 million! Many of these will be at the smaller size range, e.g. insects and micro-organisms. It is estimated that up to 50 000 species are becoming extinct each year, up to 1% of the total. To say that this loss is unsustainable and unacceptable is stating the obvious.

---

this technique it can be shown that the most likely primitive number of n is 11–13, depending on the plant family.

The genus *Rosa* is thought to have a basic haploid number of 7, because among this genus can be found species with chromosome numbers which are a multiple of 7, such as 14, 21, 28, 35, 42 and 56. This raises the question 'are polyploid offspring fertile?'. In the case of plants self-fertility would allow for the correct restoration of chromosome number after gamete production. The mechanisms involved in this are dealt with in Chapter 4. In animals this is not so easy; the probability of two polyploid individuals arising at the same time is remote, so mating of a polyploid individual with another of the same type is unlikely. Besides this there is the founder effect: our putative founders of a new species contain less variation than the whole gene pool – some characteristics may be detrimental to individuals or the species.

Interestingly, the morphological differences found between plants with different ploidy levels are not always significant. This is shown very well in the plant *Saxifraga pennsylvanica*, a North American species. In this particular species diploids, triploids and tetraploids are common, with cell size being the consistent difference between them. The polyploid species seem to be better able to cope with different habitats, being quite at home in harsher environments. It should be noted that, although polyploids with even multiples of the haploid genome are fertile, the uneven multiples (e.g. triploids) are often infertile. If a species cultivar is triploid and therefore infertile it would naturally be expected that seed production would also be limited: an ideal characteristic in a food plant whose seeds are not normally eaten.

Production of polyploid forms can be used to advantage, e.g. to obtain larger or more spectacular flowers from a species. In a simple case this might involve producing tetraploid antirrhinums from the more normal diploid varieties. With growing demand for the production of ever more spectacular floral forms, plant breeders have turned to exotic flowers such as orchids. As orchids are such an important commercial crop (over 50 000 hybrids have been cultivated) and constitute a large plant family (over 17 000 species in 750 genera), the results of ploidy changes in orchids serve well to exemplify the methods and results of ploidy changes in plants.

The principal method of production of polyploids involves the use of colchicine, (see Chapter 10 for details of artificial manipulation of genomes). This molecule binds to tubulin, the primary molecule of the spindle fibres, which control the chromosomes during cell division. Careful use of colchicine in tissue culture can sometimes, after many days, produce polyploids of various types as cells revert to interphase before cell division is completed. If this is done in sterile cultures of meristem material it is possible to produce entire tetraploid plants from the tetraploid callous material. However, this process has some drawbacks: if all the transformed cells are not of the same ploidy level then a mixoploid plant is produced that is a chimaera of different chromosome numbers in different cells. The polyploids produced in this way are generally termed autopolyploids since they contain multiple copies of their own chromosomes. Increasing chromosome numbers have a particular effect upon plant cells: cell volume also increases in approximate proportion. Figure 9.1 gives direct measurements of *Cymbidium* orchid guard cells, one from a diploid and one from a tetraploid.

Production of viable polyploids is easier in plants than in animals because in animals survival is often compromised by the physiological

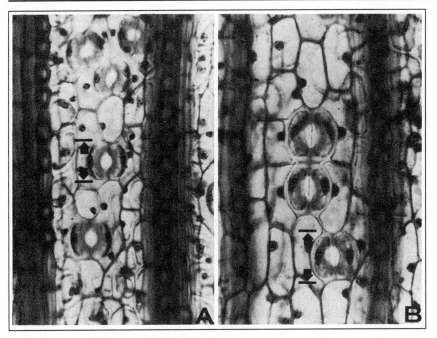

Figure 9.1 Epidermal cells from the lower surface of *Cymbidium* Peter Pan 'Greensleeves' leaves. The guard cells are paired half-moon shaped cells. A: Diploid. B: Tetraploid. The sizes of the guard cells are measured with an ocular micrometer in a microscope. The lines and arrows indicate the limits of the measurements to be taken. Tetraploid guard cells in B are about 25% larger than the original diploid in A. A and B are taken at the same magnification. Reproduced from Wimber and Watrous (1985) Artificial induction of polyploidy in orchids. *Report of the International Centenary Orchid Conference 1985*, Royal Horticultural Society.

incompetence induced by genetic imbalance. Although a complex subject for study, accurate records are kept of chromosome abnormalities in humans. In humans triploids do not survive longer than a few hours after birth and are normally aborted early in pregnancy. According to records tetraploid humans have never survived for long *in utero*.

9.1.2 Crosses between species

It would seem that any potential for evolution and speciation via polyploid production would be better suited to plants rather than animals, in which gross changes can have disastrous effects. Even in plants, speciation has generally advanced by production of allopolyploids. Changes due to autopolyploidy can be slow. Allopolyploids are also dealt with in Chapter 10. In 1928 Krapechenko produced an artificial cross from radish (*Raphanus sativa*) and cabbage (*Brassica oleracea*), both of which have a diploid number of 18. F1 hybrids have 18 chromosomes, nine from each parent, but at meiosis pairing usually fails as the chromosomes are trying to pair across species. As a result meiosis is abnormal and zygote production virtually zero. This is a typical example of infertility in an interspecific hybrid. However, if the chromosomes undergo doubling first then a set of homologous chromosomes that can pair during meiosis is formed. From these cells can be produced seeds that give rise to true-breeding plants of $2n=36$. These plants are self-fertile, but sterile when crossed with either parent. So different from both parents is this allotetraploid that it is called *Raphanobrassica*. Although of great scientific interest this very large plant inherits little crop potential from either parent; unfortunately for farmers it takes its root morphology from cabbage and leaf morphology from radish!

The normally sterile interspecific crosses are also found in animals. A well-known example is the result of crossing horses and donkeys. There are two possible crosses between horses and donkeys: the product of crossing a male horse with a female donkey is a mule, whereas the product of crossing a female horse and a male donkey is a hinny. Horses have a diploid number of 64, whereas donkeys have a diploid number of 62. Crosses have 63 chromosomes, and chromosome pairing during meiosis is apparently impossible. Occasionally, a female mule will produce a viable ovum that is the result of the entire horse or donkey chromosome complement segregating into the resultant gamete. In this case the offspring type depends upon which type of male fertilizes the ovum.

Muntjac deer show an interesting chromosomal phenomenon, that of tandem fusions of chromosomes. *Muntiacus munjak*, the Indian muntjac and *M. rooseveltorum* (Roosevelt's muntjac) both have $2n=6$, whereas *M. reevesi* (Chinese muntjac) has $2n=46$. Although there is a large disparity between the chromosome numbers, interspecific hybrids are viable because the reduction in number has been brought about by tandem fusions. Whether this is a cause of speciation or the result of chance events independent of evolution is unknown.

Wild crosses giving rise to new, viable, offspring – essentially instant speciation – are unknown in animals but are known in plants. One such case has resulted in the spread of a new grass in Europe. Two species were involved: *Spartina maritima* ($2n=60$) from Europe and *Spartina alterniflora* ($2n=62$) from North America. These plants are saltmarsh species. The American plant was introduced by accident into Europe during the nineteenth century. Over time the two species became intermixed on salt marshes. Eventually, probably as a result of abnormalities in meiosis in one of the parents, a sterile hybrid was formed. This plant has a chromosome number of 62, rather than the expected 61, which is why problems of meiosis were expected. Although both species have morphologically similar chromosomes the hybrid is sterile. This F1 cross was called *Spartina* × *townsendii*. Later on, during the 1890s, a fertile cross between two of these supposedly sterile hybrids was found. This vigorous cross was larger and more fecund than either

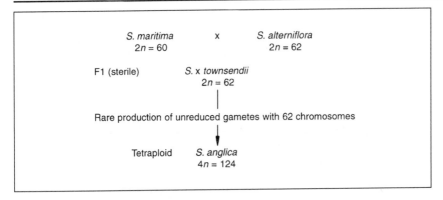

parent. It has a chromosome complement of $2n=124$, with occasional plants of $2n=122$ or $2n=120$. The new species, *Spartina anglica*, has been recreated artificially to demonstrate the origin of this species. Figure 9.2 shows the suggested path of origin of this species.

This is not the only case of natural speciation by allopolyploidy, although it is one of the best documented. By crossing two species of mint, *Galeopsis pubescens* and *Galeopsis speciosa*, each of $2n=16$, it became possible to create a viable tetraploid hybrid with 32 chromosomes. Strangely, this hybrid was found to resemble very closely a third naturally occurring species, *Galeopsis terahit*. This species also has 32 chromosomes and 16 bivalents at meiosis. It is proposed that the artificially created hybrid is essentially identical to the wild species.

9.2 Speciation by karyotype changes

Karyotype changes that result in speciation are more difficult to follow than simple changes in ploidy, but these are processes that have had a greater influence on long-term speciation and evolution. Although there are two genera in which karyotype changes have been looked at in considerable detail, other organisms have also been studied in depth. One of the two most studied genera is *Drosophila*, primarily because of its short generation times, ease of culture and the presence of polytene chromosomes. The other species that has been studied in some depth is *Homo sapiens*, in this case to satisfy curiosity.

Drosophila persimilis and *D. pseudoobscura* are difficult to distinguish morphologically, a common finding in the Insecta among closely related species, but crosses show clearly that there is a fundamental difference between these two species; the progeny are sterile males and fertile

Figure 9.3 Male karyotypes in several species of *Drosophila*. The X and Y chromosomes are at the bottom of each drawing. (1) *D. willistoni*; (2) *D. prosaltans*; (3) *D. putrida*; (4) *D. melanogaster*; (5) *D. ananassae*; (6) *D. spinofemora*; (7) *D. americana*; (8) *D. pseudo-obscura*; (9) *D. azteca*; (10) *D. affinis*; (11) *D. virilis*; (12) *D. funebris*; (13) *D. repleta*; (14) *D. montana*; (15) *D. colorata*. Reproduced from Burns and Bottino (1989) *The Science of Genetics*, Macmillan.

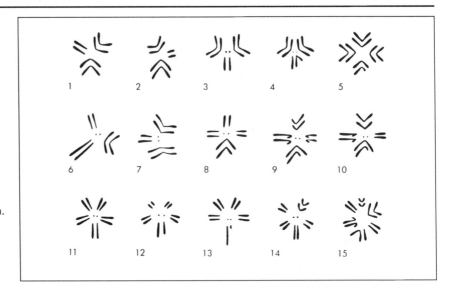

females. The karyotypes of various species of *Drosophila* (Figure 9.3) shows that it has been repeatedly rearranged by inversions and translocations.

It has long been known that closely related species often exhibited karyotypic differences, but this is not always the case. Salamanders of the genus *Desmognathus* have a diploid number of 28, although considerable variation in DNA mass between the different species has been measured. Exactly how any measured differences between species are distributed among the chromosomes is unknown, but evidence has been found for this phenomenon in other groups of species.

Species of Leguminosae plants of the genus *Lathyrus* generally have the same diploid number of 14, although constancy of chromosome number does not reflect a constancy of nuclear DNA content. Seven species with $2n=14$ were studied. They had widely differing nuclear DNA contents varying from approximately 6 pg to 30 pg. This remarkable variation in DNA content seems to be directly related to an increase in the amount of repetitive DNA. In many ways one could expect such a result: closely related species, which, it is reasonable to assume, have closely related requirements, will have closely related numbers of active genes. The differences in karyotype between species can therefore be thought of as differences in the amount of non-coding DNA sequences. The major variation in genome size (Box 9.2) is due to repeat sequences, usually packaged in heterochromatin. By taking the human genome content as 1, Table 9.1 shows clearly that there is little correlation between complexity of an organism and nuclear DNA content.

> **Box 9.2 Counting the number of genes in a genome: the example of *Homo sapiens***
>
> Before any tentative calculation of gene numbers can be made we have to have a clear idea of what we mean by genes. There are, broadly speaking, two definitions of genes. The first of these is a mappable locus that generates a mutatable phenotype. A more recent definition may be any stretch of DNA that is transcribed. This in itself raises problems, as it is necessary to know whether to include in this the controlling sequences of promoters and operators. This is an important point for the transformation of a simple DNA mass into a number of genes.
>
> In the past it has been suggested that the total number of genes in humans is between 50 000 and 100 000. Although this very wide range is too imprecise to be informative, one thing is generally agreed upon: half of all genes are either directly or indirectly associated with the construction and running of the central nervous system.
>
> The human genome is approximately 3000 Mb in size. From this we only have to decide on a value for the density of genes per megabase. This in itself is not so easy. Genes are not uniformly distributed along all the chromosomes. By using several different techniques a consensus value ranging from one gene for every 40 kb to one gene every 50 kb can be obtained. Using this range of values we can narrow our estimate of the number of genes in the human genome to between 60 000 and 75 000.

DNA content is apparently crucial to evolution. A major question is 'what factors operate to maintain this DNA diversity?'. The two major forces that operate to affect DNA content are natural selection and genetic drift. These are difficult to separate in populations because, although the processes are different, the end results are indistinguishable. A comparison can be made between sequence diversity in specific genes of related species. When this is done for *Drosophila* species, observed and expected variation due to neutral drift are sometimes very different.

It can be seen that it is quite easy to demonstrate the process of speciation when looking at the chromosomes of a particular group. A more difficult process is to infer taxonomic relationships – phylogenies – from karyotypes; deduction of phylogenies from karyotype is nearly

Table 9.1 Relative DNA content of various organisms. Data taken with permission from John and Miklos (1988)

Organism	Size of genome expressed as a fraction of the human genome
Urochordata	0.68
Hemichordata	0.20
Cephalochordata	0.20
Agnatha	0.4–0.8
Reptilia	0.5–1.5
Aardvark	1.7
Muntjac deer	0.7
Diptera	0.04–0.25
Orthoptera	0.9–3.6
Plethodon (salamander)	5–20

always done with a prior knowledge of the basic relationships to be found in a group. Within Amphibia, most taxonomy is based on morphology. This does not mean that chromosome changes are not crucial to the evolution of this group, they most obviously are, but only that they are more difficult to define. Whereas the changes in morphology follow the premise of evolution to a niche, chromosomal changes may be quite large yet only have a minimal effect on morphology. Using additional data from molecular biology studies, such as protein analysis, it has been possible to resolve the monophyletic relationships of the Amphibia as a whole and also the relationships found within some of the groups (e.g. caecilians). At the moment other groups are not nearly so well resolved. The presence of supernumerary chromosomes (Chapter 3) in some groups can be used for the investigation of relationships between closely allied species, but only with great caution. In some species of amphibian the size and number of supernumerary chromosomes varies quite widely without significant changes in the animals' morphology, even though in some cases the supernumerary chromosomes have been shown to be transcribed.

The role of chromosomes in evolution can be assumed simply because different species have different chromosome numbers and structures; such variation would be meaningless without natural selection. Without selection species would be expected to have random collections of chromosomes. In mammals, for example, $2n$ is known to vary between 6 and 84, although this may not be an exhaustive range. Within any group it is considered that the number of chromosomes a species has is dependent upon the number and form of chromosome rearrangements which the species has undergone. An example of this is shown in the Lemuridae (lemurs) of Madagascar. Robertsonian translocations (Chapter 6) have progressively reduced the chromosome number in this group, whereas in the Cercopithecinae (Old World monkeys) there have been fissions of chromosomes to increase the chromosome number.

9.3   Speciation in an ancient mammalian group

A particularly instructive group to look at is the Marsupiala of Australia and South America. There is a considerable range of chromosome numbers to be found in the Marsupiala, although not as wide a range as is to be found in placental mammals (Box 9.3).

Box 9.3   Marsupials and speciation by stealth

There are three extant lines of primary mammalian evolution;

- monotremes
- marsupials
- placentals.

The marsupials can be diffentiated from other mammals in several ways. Carrying their young in a pouch would alone give them a separate identity, but the combination of a series of discrete and unusual features gives the marsupials a unique status among mammals. These are:

1. possession of a protective pouch;
2. presence of a cloaca;
3. a female genitourinary system that shows paired uteri and a double vagina;
4. testes situated in front of the, frequently bifid, penis;
5. the replacement of only one tooth – the milk molar;
6. epipubic bones associated with the pelvic girdle;
7. a fenestrated bony palate;
8. no corpus callosum linking the right and left hemispheres.

After the early Mesozoic period there were two major land masses – Laurasia in the north and Gondwanaland in the south – separated by the Tethys Sea. This was a major barrier to the north–south spread of fauna.

The marsupials most likely originated in South America during the Cretaceous period, from where they radiated into North America and probably Europe as well. These North American species probably became extinct because of the challenge of placental mammals during the Miocene period. Pressure was not so severe in South America, and some 77 species have survived into the present. None of the modern marsupials of S. America can compete with the strange forms that the now extinct types had, such as the hyena-like *Borhyaene*, or the sabre-toothed *Thylacosmilis*.

Southwards, marsupials radiated via Antarctica to Australia, where there are now approximately 161 species remaining. Once again there has been some loss of diversity, with such forms as the lion-like *Thylacoleo* becoming extinct as well as *Sthenurus*, a kangaroo standing 3 m tall.

With no competing placental mammals the radiation of marsupials into the available ecological niches became as complete as the radiation of placental mammals in the rest of the world.

Table 9.2   The range of diploid chromosome numbers found in the marsupials of South America and Australia. Data taken with permission from Hayman *et al.* (1987)

Family	Diploid chromosome number
Didelphidae	14, 18, 22
Microbiotheriidae	14
Caenolestidae	14
Dasyuridae	14
Myrmecobiidae	14
Peramelidae	14
Thylacomidae	18, 19
Phalangeridae	14, 20
Burramyidae	14
Petauridae	10, 12, 16, 18, 20, 22
Macropodidae	10, 11, 12, 13, 14, 15, 16, 18, 20, 22, 24, 32
Vombatidae	14
Phascolarctidae	16
Tarsipedidae	24
Notoryctidae	20

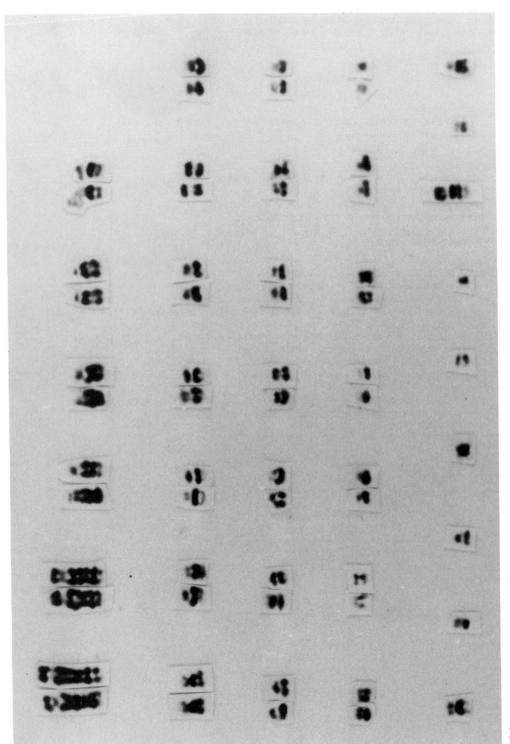

(a)

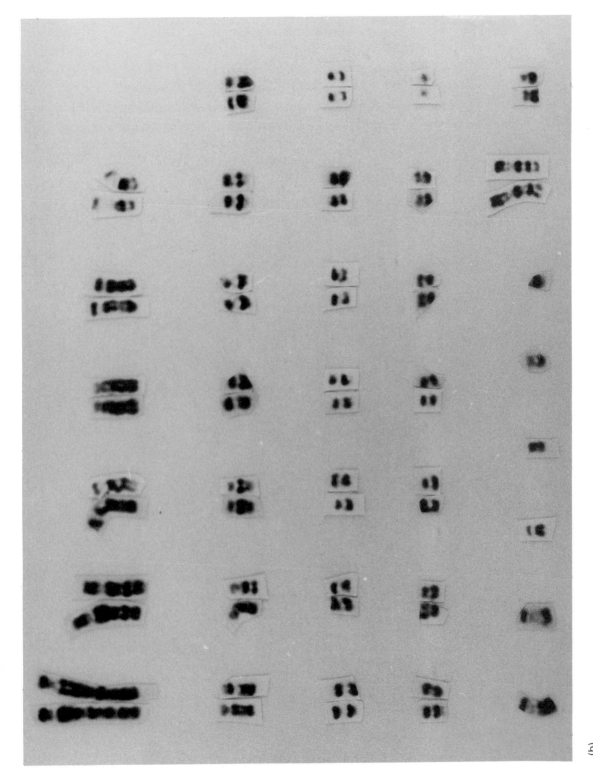

(b)

Figure 9.4 G-banded mitotic idiograms of a male and female echidna, *Tachyglossus*, showing complexity of monotreme karyotypes. (a) Male. (b) Female. Courtesy of Dr J. Watson.

Table 9.2 shows the range of chromosome numbers found within the living marsupials of two continents.

The marsupials are not the most ancient mammalian group. That is a position held by the egg-laying mammals, Monotremata. Figure 9.4 shows the karyotypes of a male and female echidna, a species of monotreme spiny anteater from Australasia. The complexity of monotreme karyotypes renders them less instructive than marsupials in explaining the role of complex karyotype changes in species evolution. Figure 9.5 is an idiogram of another monotreme, a male Australian platypus.

It has been postulated that karyotype features and changes can be described in four basic ways for all the marsupials, not just the Australian ones.

1. Species with low diploid numbers have larger chromosomes.
2. Nuclear DNA content varies from less than that of man to 1.4 times that of man.
3. The chromosome complements do not vary substantially between the two groups.
4. The commonest diploid content is $2n=14$.

Figure 9.5 G-banded mitotic chromosomes of a male *Ornithorhyncus*, duck-billed platypus, from Australia. Courtesy of Dr J. Watson.

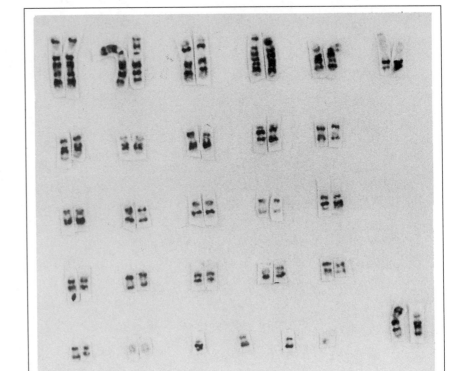

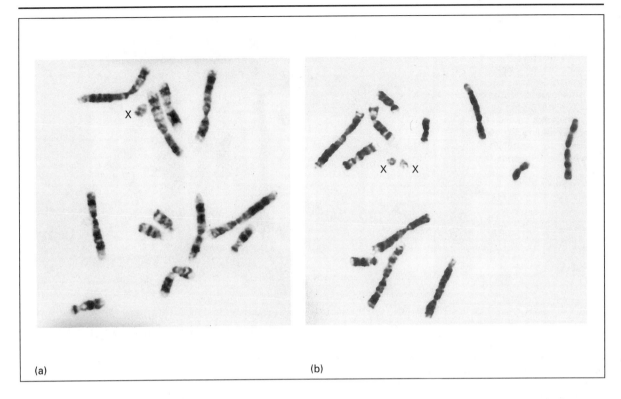

(a)   (b)

This is so for both Australian and South American species. In fact, so similar are the chromosomes between some Australian and South American species that the homology extends to G-banding as well. The striking similarity in G-banding is well demonstrated in Figure 9.6, which shows idiograms of a woolly opossum from South America and a marsupial mouse from Australia, both of which are $2n=14$. Figure 9.7 shows a solid stained preparation of a male *Sminthopsis crassicaudata* showing the presence of the extremely small Y chromosome. Such very small chromosomes are frequently lost when they are enzymically digested to reveal the G-banding pattern; this is what happened to the Y chromosome in Figure 9.6a. These data would suggest that the ancestral chromosome number was 14. Since the changes are uniform across the whole marsupial group they bear no direct relationship to the original karyotype.

The very close similarity between chromosome banding patterns within the marsupials would also imply that there is a conserved complement that runs throughout the chromosomes of marsupials and that the main changes to the karyotype have been brought about by chromosome fissions and centromere movement.

Some species have undergone a reduction in numbers of chromosomes from $2n=14$. *Wallabia bicolor*, the swamp wallaby, is just such a

Figure 9.6 G-banded mitotic preparations of two species of marsupials from different continents. The banding patterns are indistinguishable. (a) *Caluromys lanatus*, male, a woolly opossum from South America. The small Y chromosome has been lost during treatment to show G-bands. (b) *Sminthopsis crassicaudata*, female, fat-tailed marsupial mouse from Australia. Both species are of the ancestral 2n = 14 type. Courtesy of Professor D. Hayman, University of Adelaide.

Figure 9.7 Male *Sminthopsis crassicaudata*, fat-tailed marsupial mouse, $2n = 14$. Solid-stained preparation showing the extremely small Y chromosome in this species. Chromosomes this small can easily be lost during the enzyme digestion used to show G-banding. Courtesy of Professor D. Hayman, University of Adelaide.

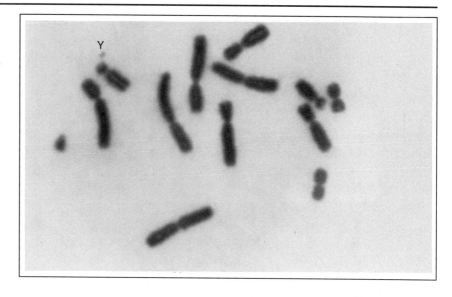

species, with the rather unusual state for a mammal of $2n=10$, XX in females and $2n=11$, XY_1Y_2 in males. This is shown in Figure 9.8.

9.3.1 Rock wallabies

A single group serves well to demonstrate that evolution and karyotype changes can run in parallel. In Australia there are a number of species

Figure 9.8 Solid-stained chromosomes spreads of *Wallabia bicolor*, the swamp wallaby. (a) Female, in which $2n = 10$, XX. (b) Male, in which $2n = 11$, XY_1Y_2. Photograph courtesy of Professor D. Hayman.

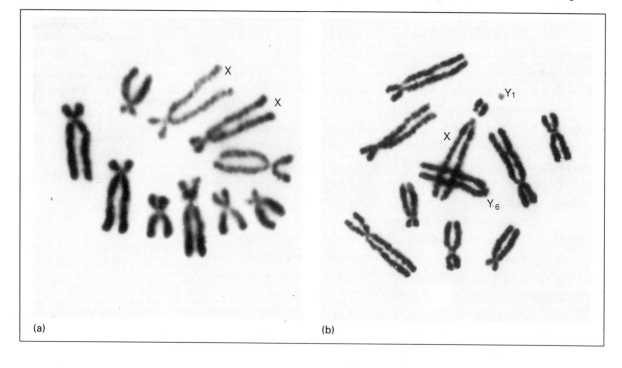

(a)　　　　(b)

and races of rock wallaby of the genus *Petrogale*. These are a particularly instructive group, as *Petrogale* is chromosomally diverse to a surprising extent. The rate of both taxonomic and karyotypic change in this group seems to vary. This may well reflect a fact that is frequently forgotten: evolution of species is not the same as evolution of karyotype.

The genus *Petrogale* is made up of approximately 15 separate species, with several more recognized races. Rock wallabies are in the family Macropodidae, comprising kangaroos and wallabies. The primitive macropodid karyotype is $2n=22$, derived from the original $2n=14$ of early marsupials.

Four species of the genus *Petrogale* – *P. persephone*, *P. xanthopus*, *P. rothschildii* and *P. lateralis*, have the primitive Macropodid chromosome number, $2n=22$. These species are scattered across Australia in isolated geographical locations.

The remaining species have undergone various changes, some more profound than others, resulting in the range of karyotypes found in this group. Figure 9.9 summarizes these changes. The natural breakpoints correspond to areas susceptible to damage *in vitro*.

We can surmise from this that, although karyotype changes may not be fundamental to the process of evolution, they will contribute to species isolation. Evidence from *Petrogale* implies that heterozygotes arising from hybrid crosses tend to be infertile, more so males than females. Isolation resulting from karyotype changes cannot be complete

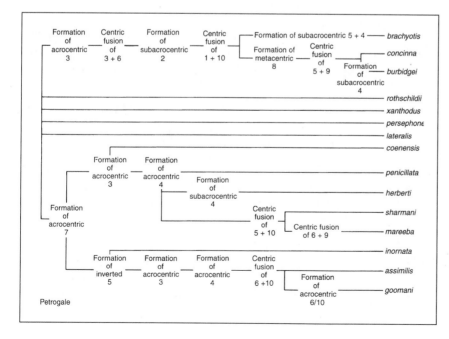

Figure 9.9 This tree shows the phylogenetic relationships between several species of rock wallaby, *Petrogale*. Formation of subacrocentric and acrocentric chromosomes is achieved by movement of the centromere. Data taken from Eldridge and Close (1993).

and immediate. To suggest that a single karyotype change originating in a population could be the cause of speciation would be wrong, since an individual, not a species, would be reproductively isolated.

Such changes as are found in karyotypes tend to be large when compared with single base changes, although they do not necessarily have such a great effect on the phenotype. For karyotype changes to be propagated through a population, they must, of course, still allow for reproduction otherwise the chromosomal changes would disappear in a single generation. Reproduction is the only way that changes can be fixed in a population; only then can speciation via karyotype changes take place. This may be with the next generation or the one after that, slowly moving the chromosomally rearranged group away from the progenitor species. *Petrogale* has provided us with an interesting example of those slow changes which gradually result in isolation and speciation.

9.4   Evolution of the hominids

Variation within species is also quite common, so it has been possible to demonstrate that in gorillas there is considerable variation in heterochromatin content. Although there is still much debate as to the exact relationships between the various members of the superfamily Hominoidea, comprising *Hylobates* (gibbons), *Symphalangus* (siamangs), *Pongo* (orang-utans), *Pan* (chimpanzees), *Gorilla* (gorillas) and *Homo* (humans), the relationships between the developing and changing karyotypes of this group are quite instructive. A simple example is that of the two races of orang-utan. They are distinguished by whether or not they have a pericentric inversion of chromosome 3. Interestingly, it is now known that the rate of molecular evolution is not consistent across different phylogenetic groups. Similarly, pericentric inversions of Y chromosomes have been seen in man and chimpanzees. Such karyotype changes as these are good examples of the problems that can arise from jumping to conclusions about speciation simply from altered chromosomes; carriers of these rearrangements freely interbreed with other members of the species.

The phylogeny of primate chromosomes, when based on chromosomal morphology, is consistent with gene mapping studies and protein structure as well as gross morphology of the species. It seems quite reasonable to deduce from this that, although morphological and behavioural changes may influence the concept of a species, it is

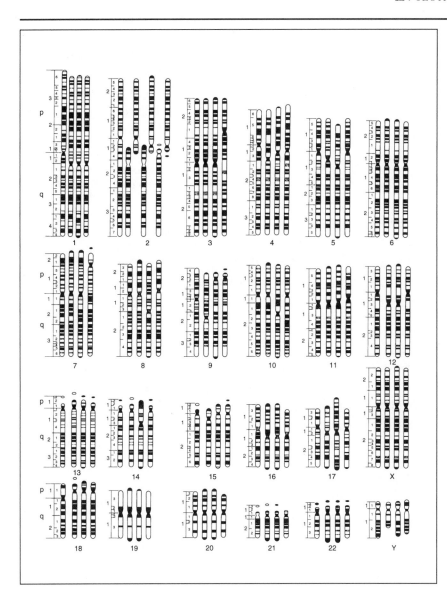

Figure 9.10 Comparative chromosome structure of man, chimpanzee, gorilla and orang-utan. This clearly shows the homology between chromosomes of these diverse species. Reproduced from Yunis and Prakash (1982).

chromosomal rearrangements that are the most robust generators of reproductive isolation. It should also be noted that the range and type of chromosomal changes that can occur are very similar in all organisms, sometimes indistinguishable between closely related species.

Figure 9.10 shows the banded karyotypes of several great apes and serves well to show that there are striking structural similarities in the chromosomes of this group. It is perhaps not so surprising that the chromosomes of this group are very similar when one considers that there are, on average, only 8.2 differences per 1000 amino acid sites between the proteins of man and chimps and that there are many types

Table 9.3 The broad changes in the hominid karyotype. Data taken with permission from Dutrillaux and Couturier (1981) and De Grouchy (1987)

Chromosome 7	Through primate evolution this has been involved in more than 20 rearrangements
Chromosomes 6, 14 and 15	These formed in the common trunk of the orang-utan and modern man
Chromosomes 3, 10, 11 and 20	These formed after the divergence of the orang-utan
Chromosomes 4, 5, 8, 12 and 16	These formed as at present before the separation of the macaques, baboons, and rhesus monkeys
Chromosomes 19 and X	These appeared after the Old and New World monkeys diverged
Chromosomes 13, 21 and 22	These are so stable that they probably appeared in a common mammalian ancestor as they are virtually identical throughout the primates

of DNA that when hybridized to related species of apes show gross similarities.

In the case of primates there is an apparent slowdown in molecular evolution. This is thought of as being due to a progressively lengthening generation time. It is worth getting this rate of evolution in perspective: 10^6 human generations will take about 15 million years at 15 years per

Figure 9.11 Evolution of the hominids.

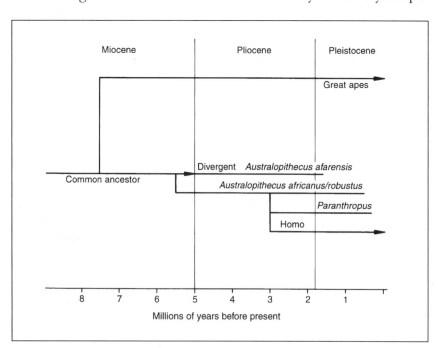

generation. It would take mice, with a generation time of 3 months, only 250 000 years to achieve this number of generations. Even faster would be the same number of generations in a bacterium: 76 years. The implications of this are obvious: it will take much longer for the effects of evolution to become apparent in species with long generation times.

Table 9.3 details the broad changes found in the evolution of the human karyotype, while Figure 9.11 shows the time scale of divergent hominid species.

Although virtually any phylogenetic investigation will result in questions regarding the relationships between chromosomes, one of the most intriguing biological puzzles is the development of the X and Y chromosomes (Chapter 8). By looking at the changes in X and Y chromosomes in *Drosophila* and the primitive mammals (monotremes and marsupials), it has been suggested that the differentiation between these two chromosomes has originated by slow changes, gradual reduction of the euchromatin being followed to a lesser degree by accumulation of heterochromatin (Chapter 2). The remaining question 'why accumulation of heterochromatin?' simply cannot be answered.

10 Artificial manipulation of genomes

Summary

Recently a considerable amount of media attention has been devoted to the subject of 'genetic engineering and biotechnology' and the way in which it has the potential to alter our lives through gene therapy of genetic disease and increased food production. The focus of research and press interest has been the addition or alteration of one gene in an organism, for example the Flavr Savr™ (non-ripening) tomato or cystic fibrosis gene therapy. However, humans have been manipulating genetics for hundreds of years via breeding and selection experiments. Also in nature, genetics is a dynamic process with hybridization and mutational events occurring continuously and eventually producing new species.

The essential difference between natural genetic change, traditional genetic manipulation and current genetic techniques is that we now have the ability to combine characteristics of species that were previously incompatible. There is also heavy commercial involvement.

This chapter examines some of the possibilities using both conventional breeding strategies and the newer technologies. Whole chromosome sets can be manipulated to produce haploids (n = half the normal number of somatic chromosomes) and polyploids, in which the somatic chromosome number is increased by at least one whole set, e.g. triploids (3x), tetraploids (4x), etc. On a more detailed level, manipulation of single chromosomes can be used to produce alien substitution and alien addition lines in crop plants. Such manipulations can often be achieved using conventional breeding techniques and involve large-scale alteration of DNA amounts. Genome manipulation, which can be used to produce a variety of chromosomal and DNA effects often not achievable with conventional breeding, includes somatic fusion and somaclonal variation. These techniques operate to produce subtle genic changes; alteration of chromosome number is generally not the aim. Finally, at the most intricate level, single-gene transfer technology is the only technique discussed that involves manipulation of mammalian genomes.

10.1 Manipulation of whole chromosome sets

Most organisms are intolerant of variants in chromosome number, let alone reproduction or loss of whole chromosome sets. In general, mammalian genomes are more sensitive to the manipulation of whole chromosome sets. Genome manipulation in animals causes severe physiological disturbances, complications in sex determination (X–autosome ratios and dosage balance), rejection of the polyploid fetus by the diploid mother *in utero* and finally the complication of obligatory sexual reproduction, i.e. since there is no means of asexual propagation the polyploid species cannot be maintained.

The usual experimental organisms involved in whole genome manipulation are plants, fish and shellfish. This is almost certainly partly the result of the ability to culture the manipulated organisms outside the parental environment and may well be related to natural ploidy levels. Further advantages of using these organisms are listed below.

Plant genomes seem to be far more plastic than those of animals: many plants are polyploid, can tolerate huge percentages of repetitive ('junk') DNA and are also highly responsive to external stimuli such as changes in the environment. Plants also have the innate ability to regenerate whole plants from single cells (totipotency), thus allowing manipulation to be carried out at many different cellular levels. After manipulation, the 'plant' (cell, tissue section or organ) can be aided in its development by *in vitro* tissue culture, considerably improving its chances of survival. Conditions are very specific to each variety and vary with the age of donor plant and type of tissue used. The optimization of the tissue culture conditions is vital to success and can greatly affect the final outcome. Often several different types of media are required at different stages to ultimately successfully regenerate a whole plant. It is a technique that is often described as more of an art than a science.

For most species of fish and shellfish, eggs are readily available in massive quantities, can be fertilized under specifically controlled conditions and do not need to be returned to the reproductive tract of the mother for completion of development (as is the case with mammals). Many are able to tolerate induction to triploids or tetraploids. This plasticity may arise because many fish, although classified as diploid, are hypothesized to actually be tetraploid, having undergone radical genome evolution from diploid ancestors.

In general, whole-genome manipulation can be divided into two categories.

1. Production of haploids: reduction of the normal diploid constitution of an organism to its gametic chromosome number (n).
2. Production of polyploid:

   • autopolyploids: an increase to a higher multiple of the same chromosome set to produce autotriploids or autotetraploids;
   • allopolyploids: the combination of two distinct species with a consequent increase in basic chromosome number over the original parents.

16.1.1 Haploid production (Box 10.1)

Examples are restricted to plants, because there is little, if any, commercial demand for haploid animals and also they do not survive! Attempts have been made to produce haploid zebrafish for developmental genetic studies, but survival is rare (and short).

Haploid plants are useful for two main reasons.

1. Homozygous lines can be produced more rapidly than using traditional methods. The production of the haploid is followed by chromosome doubling (this either occurs spontaneously or is

---

Box 10.1 Production of haploids

Haploid plants can occur naturally in some crop plants, as a result of irradiation, chemical agents or certain genetic systems. The best-known system is barley: *Hordeum bulbosum*, a wild species, can be crossed with the commercial barley varieties. A genetic incompatibility system operates such that all of the *H. bulbosum* chromosomes are rapidly eliminated from the hybrid, leaving a haploid plant.

However, these systems are rare and haploids are more usually produced by the regeneration of cultured anthers or immature pollen grains (the mid-uninucleate stage, before the microspore nucleus divides to form the generative and vegetative nuclei). It is possible to culture ovules, but anthers are more readily accessible, with no requirement for dissection, and are far more numerous, producing thousands of plants from one anther. For example, barley has over 10^5 microspores per spike, all with the potential of producing a haploid plant.

---

induced by the use of colchicine) to produce a viable diploid. Genetically the resulting doubled haploids are identical to homozygous lines produced by conventional breeding.

Homozygous plants are achieved within 2 years of the initial hybrid cross, but additional years are required to bulk up the seed before replicated selection trials can take place, approximately 5 years in total. However, with conventional breeding over 5 years, plants selected for the preliminary field trial theoretically have 12% of the plant genes still segregating. On average, a minimum saving of 2 years over traditional culture methods is possible with haploid cereal crops.

2. Detection and selection of recessive mutants is possible. With the selection of mutants the reduction of a plant to the haploid level and then doubling up of the chromosome number allows expression of recessive genes previously 'hidden' by dominant alleles. Thus, fewer plants need to be tested for recessive characteristics. Mutagenesis at the haploid level is also more efficient as only one locus needs to be hit before doubling. With diploids the same effect only occurs if a double mutation occurs at the same locus: a very rare event. Biochemical selection such as the culture of embryos on medium containing bacterial toxins can select for disease resistance and is also more effective at the haploid level, for similar reasons.

Many of the plants obtained from anther culture are spontaneously doubled haploids. Aneuploids may occur as a result of random loss of chromosomes, an effect that often depends on the culture conditions. A particular problem with the Gramineae (a family which includes wheat, barley, oats, etc.) is the production of albino variants.

One of the first commercial releases of a doubled haploid was

'Mingo' barley by Ciba-Geigy Seeds Ltd. The cross for Mingo was made in 1974, and by 1979 it became a new licensed variety. This was produced using the *Hordeum bulbosum* genetic system, however anther culture is now the preferred method for haploid production in barley.

Apart from barley (which, of the cereals, seems particularly well suited to haploid production), anther culture conditions have only recently been refined for most of the major crops. For example, doubled haploid varieties of rice, wheat and tobacco are now in commercial production. Biochemical selection has been successfully used to produce disease-resistant tobacco and rape varieties and salt tolerance and herbicide resistance in tobacco. Even trees have been produced via anther culture, e.g. apple cv. Delicious and horse-chestnut. The regeneration of trees, in particular, is hampered by the very short annual season during which they flower and produce suitable experimental material. This is compounded by their long life cycles, which do not make for quick screening of regenerants (even so, manipulation using anther culture is still much faster than conventional breeding).

10.1.2   Production of polyploids

Production of polyploids is also mostly restricted to plants, fish and shellfish. Their ability to tolerate ploidy manipulation has already been discussed, but the reasons behind the desire to induce polyploid individuals in each varies. For a list of some commercially important polyploids see Table 10.1.

Table 10.1 Commercially important polyploids

			Cross that produced polyploids	Genomes
Autopolyploids				
Beta vulgaris	Sugar beet	$2n = 3x = 27$		
Malus pumila	Apple	$2n = 3x = 51$		
Solanum tuberosum	Potato	$2n = 4x = 48$		
Oncorhynchus mykiss	Rainbow trout	$2n = 3x = 87$		
Crassostrea spp.	Oyster	$2n = 3x = 30$		
Allopolyploids				
Triticum aestivum	Wheat	$2n = 6x = 42$	*Triticum monococcum* × *T. speltoides* × *Aegilops squarrosa*	AABBDD
Nicotiana tobacum	Tobacco	$2n = 4x = 48$	*Nicotiana sylvestris* × *N tomentosiformis*	SSTT
Triticale		$2n = 8x = 56$	*Triticum aestivum* × *Secale cereale*	AABBDDRR
		$2n = 6x = 42$		AABBRR
Musa spp.	Banana	$2n = 3x = 33$	*Musa acuminata* × *M. balbisiana*	Varies
Eleusine coracana	Millet	$2n = 4x = 36$	Unknown wild *Eleusine* species, possibly *E. indica*	?
Gossypium hirsutum	Cotton	$2n = 4x = 52$	*Gossypium herbaceum* × *G. raimondii*	AADD

> **Box 10.2   Production of autopolyploid plants**
> Triploid plants are generally produced by crossing tetraploids (4x) with diploids (2x). Tetraploids can be produced using colchicine (Chapter 1) to induce chromosome doubling. Polyploid plants have the advantage that they can easily be clonally propagated once the initial stock has been produced. Some reproduce via tubers and bulbils, others set seed; grafting and micropropagation can also be used. Hence the initial production is a one-off event. With clonal propagation all offspring resemble the parental stock, i.e. there is uniformity of product.

Autopolyploids

Plants (Box 10.2)

The reasons for producing autopolyploid plants include the following.

1. Increase in genetic variation. The effect of polyploidy varies tremendously, depending on the genome combinations already present. Multiplying even the same basic chromosome sets already present within a plant can be beneficial. Increasing the chromosome number produces an increase in the amount of genetic variation present in the plant. For example, a diploid heterozygous at a gene locus has the following possible constitutions:

Parental	Gametes
AA	A
Aa	1A:1a
aa	a

    A tetraploid has the following possible constitutions:

Parental	Gametes
AAAA	AA
AAAa	1AA:1Aa
AAaa	1AA:4Aa:1aa
Aaaa	1Aa:1aa
aaaa	aa

    There is a predominance of parental heterozygous types and the occurrence of several types of heterozygote in the offspring due to random segregation. Although the 'A' effects may dominate, there will invariably be subtle differences produced by the dosage effects of the four genes, therefore the offspring do not necessarily closely resemble the parents. In general, these are termed 'Gigas effects', the offspring being morphologically larger with thicker leaves and respond differently to environmental conditions.

2. Sterility. Odd-numbered autoploids are generally sterile. Thus, they produce no seeds and are therefore more commercially acceptable

to the general public. Banana (triploid) is a classic example of this. Even-numbered autoploids are generally associated with infertility problems as all sets of chromosomes can pair, producing multivalent formation at meiosis (see Chapter 4 for more details on

---

Box 10.3   Production of polyploid fish/shellfish (Figure 10.1)

Triploids can be produced in either of two ways.

1. Crossing a tetraploid with a diploid. Experiments in fish indicate that crossing a tetraploid with a diploid produces better triploids than manipulation of diploid embryos, but there are technical problems in reliably producing and maintaining tetraploid fish. This technique has not proved feasible with shellfish for similar reasons.

2. Meiotic shock producing chromosome doubling during either meiosis I, meiosis II or first cleavage after fertilization.

In molluscan shellfish any of these stages can be targeted because the eggs are spawned before or just at the metaphase I stage of meiosis. At this point development is halted until the eggs are activated by sperm, enabling precise control of the start of the maturation divisions.

Fish and crustacean shellfish are less amenable to this type of manipulation. With fish, these divisions take place in the ovary before egg release and most commercial crustacean shellfish do not release their eggs into the sea, but brood their eggs and embryos.

Methods to induce triploidy in shellfish include heat (25–38°C), cold (0–5°C), the chemical cytochalasin B (a fungal metabolite) and pressure (6000–8000 psi for 10 min). Of these, the most reliable and often used is the chemical cytochalasin B. Why and how these particular environmental stresses should cause meiotic shock is not precisely known.

Heat treatment (39–42°C) is the preferred method for inducing triploid fish.

With meiotic shock treatments, manipulations have to be made each time triploids are required, and this process is not 100% effective. This currently limits the commercial uses for most species.

Figure 10.1 Ploidy manipulation in shellfish. Reproduced from Beaumont (1994) *Genetics and Evolution of Aquatic Organisms*, Chapman & Hall. (a–d) Diagrammatic representation of the maturation divisions, syngamy and first cleavage in bivalve molluscs. For simplicity only one pair of chromosomes is shown. (a) Egg at release at metaphase of meiosis I, activation by sperm; (b) meiosis I complete, first polar body extruded, sperm nucleus (male pronucleus) has entered egg; (c) meiosis II completed, second polar body extruded, male and female pronuclei unite on first cleavage spindle; (d) first cleavage. (e–k) Consequences of ploidy manipulation. (e) Shock administered at meiosis I, first polar body suppressed; (f) normal meiosis II, second polar body extruded, female pronucleus (2N) unites with male pronucleus (N); (g) first cleavage (meiosis I triploid); (h) shock administered during meiosis II, no second polar body produced; (i) first cleavage (meiosis II triploid); (j) shock administered at first cleavage; (k) tetraploid.

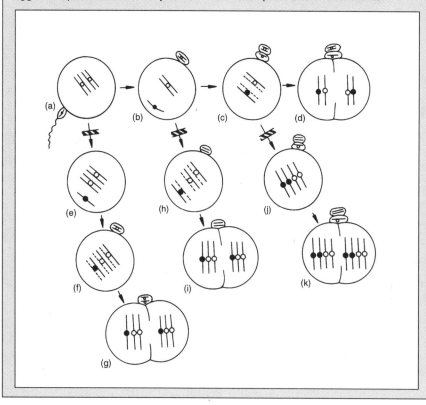

nuclear division of polyploids). Tetraploid crops are exploited more for the Gigas effects and the increase in release of genetic variation. As they are partially fertile, they are often used in the first stages of triploid production.

Some of our more important crop plants are autopolyploids. For example sugar beet (*Beta vulgaris*) is an autotriploid, as is the dessert banana, and the forage crops Italian rye grass (*Lolium italicum*) and red clover (*Trifolium pratense*) are autotetraploids.

Aquaculture (Box 10.3)

Recent interesting developments in autopolypoidy centre around the possibility of producing triploid fish and shellfish. The main reasons for the development of triploids in aquaculture are as follows.

1. Sterility. Triploid fish and shellfish are effectively sterile and therefore do not waste energy on the production of gametes. This energy is available to be channelled into a faster growth rate, i.e. to produce bigger fish. Sterility also means that non-native species can be introduced into an area without fear of an accidental release, potentially leading to interbreeding.
2. Taste. In shellfish such as oysters, the gonads spread throughout the somatic tissue, often to the extent that mature animals are unmarketable. This gonad production requires glycogen as an energy source, the depletion of which apparently has an adverse effect on the flavour according to taste trial panels. The quality of the flesh also deteriorates when they mature and spawn, restricting the selling season.

As the methods for producing triploid fish and shellfish are not 100% effective, various techniques for confirming the triploid nature of the organisms have been developed. These include chromosome and polar body counts, electrophoresis, flow cytometry (Chapter 7), microfluorimetry and nuclear sizing, principal criterion for a successful sampling technique being the ability to sample accurately a small number of cells without irreversible harm or destruction of the individual under test. Nuclear sizing (triploid nuclei are 1.5 times larger than those of diploids) has been used successfully with fish and is being further developed for shellfish.

Triploids have been successfully produced in oysters, scallops, mussels, clams and Pacific abalone. So far, only the commercial production of triploid oysters has been set up and is a routine operation. The method

Box 10.4   Production of allopolyploids

Even if fertilization of the initial cross is successful, the F1 products are frequently sterile because the differences between the two chromosome sets are often too great to allow efficient pairing and recombination at meiosis. Treatment with colchicine can double the chromosome number to the combined level of both parents, resulting in two sets of chromosomes that do not pair interspecifically (Figure 10.2).

This polyploid is fertile and can be utilized commercially. These are often called amphidiploids (as the chromosomes effectively act as diploids). Because the two sets of chromosomes are structurally and genetically different, uniform gametes are produced, i.e. one set from each parent, so the offspring are immediately true breeding and likely to resemble the parental polyploid. Where complete incompatibility exists between species, somatic fusion may be used.

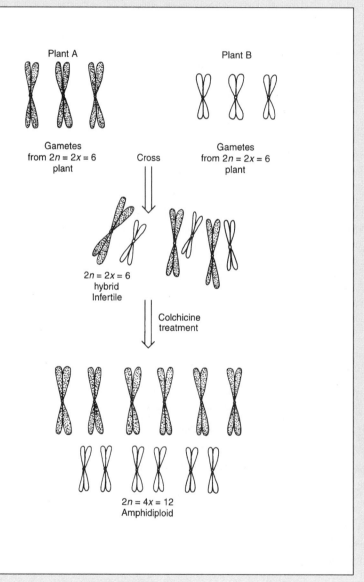

Figure 10.2   The basic stages of amphidiploid production. For further details see text.

Plant A

Plant B

Gametes from $2n = 2x = 6$ plant

Cross

Gametes from $2n = 2x = 6$ plant

$2n = 2x = 6$ hybrid Infertile

Colchicine treatment

$2n = 4x = 12$ Amphidiploid

of triploid oyster production has been the subject of a patent application (not surprising considering the high commercial value of oysters).

In fish, triploids of common carp and rainbow trout have been successfully produced by using heat shock. These fish are effectively sterile, although triploid males do exhibit secondary maturation characteristics and occasionally produce a few aneuploid spermatozoa. Maturation in females is almost completely suppressed, hence they are the preferred sex for triploid production. Commercial production of triploid all-female rainbow trout is now under way.

Allopolyploids (Box 10.4)

Mixed genomes can be desirable for their combination of useful characteristics from different varieties. If these are compatible and the different sets of chromosomes pair effectively at meiosis, crosses can be made conventionally and then used in a backcrossing programme to produce homozygous inbred lines.

Amphidiploids (Box 10.4) have sometimes been shown to be more stable than the original F1 hybrid even if the diploid F1 is fertile. For example, hybrid ryegrass (4x) is a combination of perennial and Italian ryegrasses. The resulting hybrid possesses the rapid early growth and high nutritional value of Italian ryegrass and the persistence of the perennial ryegrass. The interspecific diploid is viable, but the amphidiploid is more stable in subsequent breeding programmes.

In general, crosses are made between plants of the same species to minimize genetic incompatibility. One attempt has been made to cross two different species in an attempt to synthesize a new crop. This is triticale, a hybrid between wheat and rye with the aim of producing a novel grain and forage crop. It generally has a higher protein content, especially of the essential amino acids, than wheat. Two main polyploids have been produced $2n=6x=42$ (4x wheat, 2x rye) and $2n=8x=56$ (6x wheat and 2x rye), of which the hexaploids are the more stable. The first triticale variety, called Rosner, was released in the early 1970s in Canada. Several more commercial varieties have been produced since.

Allopolyploids include many of our common crop plants. The classic example is bread wheat (*Triticum aestivum*) with an AABBDD constitution. Other important commercial crops include millet ($2n=4x=36$) and cotton ($2n=4x=52$), all of which have been derived from wild diploid ancestors. Banana, mentioned above, also falls into this category, as the important African crops, plantain and cooking bananas, are related to dessert bananas (AAA), but contain a mixture of A and B genomes.

Box 10.5 Production of alien addition lines (Figure 10.3)

A wheat–rye amphidiploid (*Tritosecale*) is constructed and back-crossed to the wheat parent. The hybrid has a diploid wheat chromosome complement, but only a haploid number of rye chromosomes (with no pairing partners). These are gradually lost in cell division as univalents until only one rye chromosome remains while the whole complement of wheat is retained. The result is a monosomic addition line (full complement of wheat chromosomes plus one rye chromosome). If this is self-pollinated, a disomic addition line (full complement of wheat chromosomes plus two rye chromosomes) is produced.

Figure 10.3 The basic stages in the production of an alien addition line.

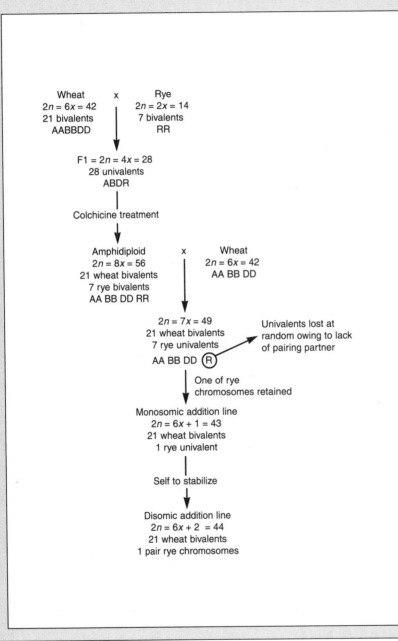

10.2 Manipulation of single chromosomes

This involves the production of alien chromosome substitution lines via aneuploid stocks and intermediate production of alien addition lines. It is used where the genetic variation required is not available in the same species and also for transfer of quantitative (polygenic) characters. The examples used to explain the process are restricted to wheat, but other systems will be discussed.

To create alien chromosome substitution lines a single pair of chromosomes from a donor variety is transferred into the recipient variety, where it replaces a recipient pair. Chromosome substitution donors must be able to compensate for the loss of the recipient chromosome species and to become integrated into the genotype without disturbing meiotic stability and fertility. Suitable donor varieties for wheat include a large number of wild relatives such as *Agropyron*, *Aegilops*, *Haynaldia* and *Elymus*, plus rye and barley.

Chromosome substitution is a multistage process, the first of which is the production of alien addition lines (Box 10.5).

Alien addition lines have been successfully produced in several crops such as wheat, oats, cotton, maize, rice, tobacco, and the brassicas. They have not proved commercially successful as they are generally unstable and undesirable traits have been carried over with the alien chromosomes. Their potential use lies in the production of alien substitution lines (Box 10.6).

The amount of alien genetic information is often too extensive to maintain the genotypic balance and attempts can be made to limit the substitution to translocations or small segments by X-irradiating the donor. These small segments are difficult to detect in the wheat genome using standard banding techniques. Routine screening is now carried out using genomic *in situ* hybridization (GISH) (Figure 10.5).

An alternative approach to the integration of small segments in substitution lines is the manipulation of the pairing control mechanisms of wheat. Chromosome 5B of wheat contains a gene, *Ph*, which causes regular bivalent formation. Deletion of this gene allows homoeologous pairing. Hence, in a wheat–alien addition line, pairing and recombination can occur between wheat and the alien chromosomes, integrating alien segments into the wheat genome.

Substitution lines have been produced in oats, tobacco and cotton, but only wheat has proved to be a commercial success so far. The problems in the other species arise when the alien chromosomes do

Box 10.6   Production of alien substitution lines

A wheat monosomic (full complement of wheat chromosomes minus
one) line is crossed with a disomic
addition line (wheat plus two extra
chromosomes), the resultant F1 has
some wheat plants which are monosomic for one of its chromosomes
with an extra alien chromosome. To
replace the wheat monosomic completely with an alien pair, the F1 is
selfed and the progeny screened for
substitution of a whole pair of wheat
chromosomes (Figure 10.4).

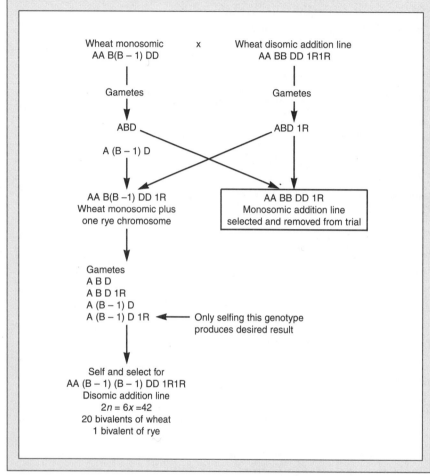

Figure 10.4 The basic
stages in the production of a
chromosome substitution
line. Note that not all
gametes will produce the
desired final result.

not fully compensate for the loss of one of the normal complement,
and also undesirable traits may be introduced along with the selected
gene.

Commercial wheat varieties such as Orlando, Zorba and Clement
carry an alien chromosome pair, a 1R (rye) for a 1B (wheat) substitution
with a 1BL/1RS translocation. The genes for stem rust resistance from
Agropyron elongatum and leaf rust resistance from *Aegilops umbellulata* have
also been incorporated into commercial varieties.

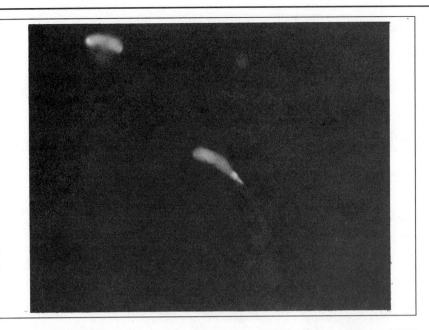

Figure 10.5 *In situ* hybridization used to detect an alien chromosome segment in wheat. Courtesy of Dr I. King and Dr T. Miller.

10.3 Manipulation producing variable genome changes

This section reviews two manipulation techniques whose outcome cannot be precisely controlled. Having said this, they can be very useful under certain circumstances. Both are applied to plants and in general do not (intentionally) involve an alteration in ploidy level, the aim being to produce more subtle genic changes.

10.3.1 Somatic fusion

Much emphasis these days is placed on the maintenance of biodiversity: the conservation of genetic stocks of ancestral species of our domesticated crops. These wild relatives often contain genes for useful characteristics, such as disease and pest resistance, drought and salt tolerance, which breeders would like to incorporate into current popular varieties. However, wild relatives are often sexually incompatible with their domesticated relatives. This may be because of evolutionary divergence between the chromosomes, different ploidy levels (many domesticated crops are polyploid with diploid ancestors) or incompatibility factors, such as cytoplasmic male sterility. Thus introgression of useful characters is not available by conventional means. To be able to incorporate these traits within our existing crops

Box 10.7   Somatic hybrid production

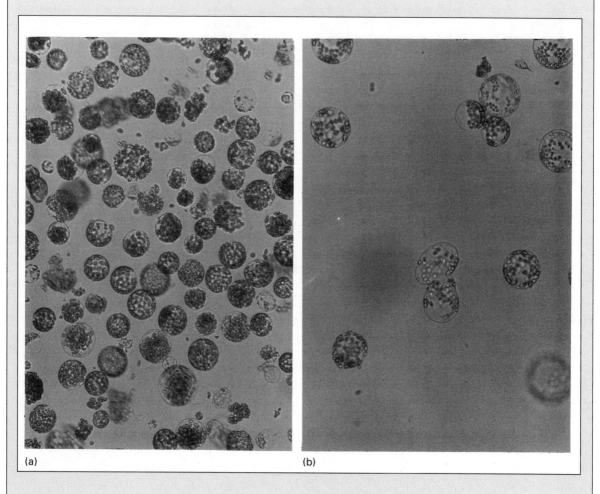

(a)   (b)

Figure 10.6   (a) Isolated mesophyll protoplasts from the somatic hybrids between *Solanum tuberosum* and *S. brevidens.* (b) Dividing protoplasts. Courtesy of Professor E. Pehu and Dr Y.-S. Xu.

Somatic hybrids are obtained by fusing protoplasts from two different species. Protoplasts are plant cells that have had their cell wall enzymically removed. They are kept in a solution of high osmotic potential, equivalent to that inside the cell to ensure that the cells do not burst when the wall is removed (Figure 10.6).

These protoplasts can then either be plated out onto culture medium to regenerate plants directly (the selection of plants which can regenerate from protoplasts is the first stage of

somatic fusion), to be used for micromanipulation or in fusion experiments, as described here.

In order to fuse two protoplasts, the membranes must be in very close contact. These tend to have a surface charge which causes repulsion. Two methods are used to overcome this.

1. Chemicals. These can be used to modify surface charges, the most commonly used being polyethylene glycol (PEG). This works by altering the distribution of pro-

teins in contacting membranes, creating lipid-rich domains that destabilize, causing fusion.

2. Electrofusion. Protoplasts are placed in a medium of low conductivity between two electrodes and a high-frequency alternating field is applied across them. The surface charge on the protoplasts becomes polarized and they migrate to the electrodes. During migration, they contact other protoplasts to form chains. When this occurs, one or more short

direct current pulses are applied. This causes reversible membrane breakdown and contacting membranes may fuse. The careful choice of pulse parameters can minimize the possibility of mass fusions, producing relatively high 1:1 fusion frequencies. The fusion products are then regenerated to whole plants using a variety of tissue culture techniques. Selection procedures for somatic fusions need careful consideration to be able to cut down on the number of hybrids screened. Even so, hundreds to thousands of hybrids may need to be evaluated to obtain the desired result. Fortunately, this is usually carried out at the tissue culture level to cut down the amount of valuable space required for a field trial.

an alternative method of crossing is required: somatic fusion (Box 10.7).

Unlike conventional crossing, this technique can be manipulated to produce several possible types of fusion events.

1. The combination of two complete genomes.
2. Partial genome transfer to produce asymmetric hybrids. One of the genomes (the donor) is irradiated to shatter the chromosomes and combined with a whole recipient genome by fusion. Most of the irradiated genome will be lost. The result is a regenerant with the vast majority of one genome and segments of another (rather like radiation cell hybrids). The reason for doing this is that wild species, although they contain useful genes for disease resistance, also contain other deleterious genes which are passed on with whole-genome somatic hybrids. Differences in chromosome constitution can also cause pairing and fertility problems. Transferring a small part of the wild genome minimizes this. It also reduces the number of backcrosses required to return the hybrid genome to as close to the cultivated parental as possible with the inclusion of the new gene.
3. The fusion of a nuclear genome from one donor and the cytoplasm of another produces cybrids. This is an important technique, as the organelles (chloroplasts and mitochondria) often contain characters of interest such as herbicide resistance factors or cytoplasmic male sterility determiners.

Protoplasts have been produced and subsequent regeneration of plants obtained from a whole range of economically important crop species such as tomato, rice, wheat, maize, oilseed rape, potato and sugarbeet. Successful fusion products are rarer. One of the major problems is that production of the protoplasts and the effect of the fusion process is a destabilizing influence on the genome. Products are often aneuploid or, even if euploid, subject to random DNA mutations (an effect called somaclonal variation).

Figure 10.7 Somatic hybrids between *Solanum tuberosum* and *S. brevidens* 84140. (a) Parent $2n = 4x = 48$. (b) Regenerant B-36. (c) Regenerant B-11. Note that the morphology of the two regenerants is different. Courtesy of Professor E. Pehu and Dr Y.-S. Xu.

This heterogeneity of product often means that the somatic hybrid itself will not be commercially viable. However, an important use of somatic hybrids is in providing new breeding material, incorporating useful characteristics from wild relatives into varieties which are compatible with and able to cross with domesticated crops, thus allowing introgression of what would have been previously unavailable characters.

The potato plants in Figure 10.7 were part of a large project to produce fusion hybrids from *Solanum brevidens* (a wild potato species) resistant to potato leaf roll virus (PLRV) and potato virus Y (PVY) and

Figure 10.8 Hybrid produced by somatic fusion of navel orange and *Troyer citrange*. All photos show the navel orange on the left, hybrid in the middle and *Troyer citrange* on the right. (a) Leaf morphology. (b) Flower morphology. (c) Cut fruits; note extremely thick skin of hybrid. Reproduced from Ohgawara *et al.* (1991), *Theoretical and Applied Genetics*, **81**, 142–143.

(a)

(b)

(c)

the commercial potato, *Solanum tuberosum* (susceptible to both viruses and incompatible with *S. brevidens*). Some fusion hybrids are resistant to both viruses and also female fertile, so they can be used in breeding trials with other commercial potato varieties to allow introgression of virus resistance into potato germplasm.

Another potential use of fusion products is in the production of commercial triploids, e.g. fusion of a navel orange with a related species, *Troyer citrange*. The hybrids produced are amphidiploids and are of no commercial value, having very thick rinds, but they can be used as breeding material for the production of triploids (Figure 10.8).

10.3.2 Somaclonal variation (Box 10.8)

Somaclonal variation is the name given to the general phenomenon of increased variation (at levels higher than those associated with spontaneous mutation) produced after the regeneration of a plant via *in vitro* culture after passage through a disorganized callous phase of growth. When somaclonal variation was first discovered, it was presented as a novel source of variability for crop improvement.

The frequency of occurrence of somaclonal variation is highly variable. Experiments on soyabean indicate 0–4 mutations per plant, whereas maize *in vitro* culture produces 0.5–1.3 variations per plant. True estimates are difficult to obtain as different groups measure these changes by different methods and at different levels of organization (morphologically, chromosomally, molecularly, etc.).

What is certain is that the effects are not strictly reproducible and are subject to batch effects, indicating that variation is both random and unpredictable. As a result, there is no guarantee that mutations of specific traits will occur. Several factors are important in influencing the rate of somaclonal variation.

1. Cultivar type.
2. Ploidy level (polyploids, because of the complex and extensive nature of their genomes, are far more tolerant of increased variation, even in the form of extra chromosomes and therefore more susceptible to somaclonal variation).
3. Tissue source and type of regeneration system, e.g. cultured leaf pieces or protoplasts (studies of protoplast cytoskeletons indicate abnormalities in microtubule organization, a prerequisite for maintaining chromosomal domains and normal division, which seem to arise directly from the method of isolation).

Box 10.8 Production of somaclonal variants

Damage of a plant tissue usually results in the initiation of cell division at or near the cut surface. If this dividing tissue is removed and cultured a callus forms, from which it is possible to regenerate whole new plants. This process can be mimicked by culturing small pieces of plant tissue such as leaf discs. Tissues such as immature embryos, influorescences and protoplasts can also be regenerated into plants and usually pass through this callus phase (Fig 10.9).

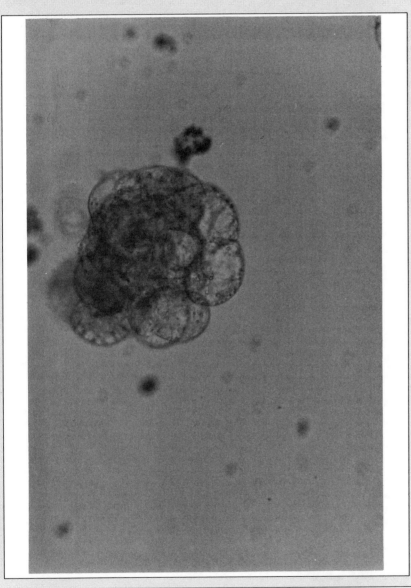

Figure 10.9
Protoplast-derived callus. Courtesy of Professor E. Pehu and Dr Y.-S. Xu.

4. Composition of culture media (the auxin 2, 4-D, in particular, has been shown to have a mutagenic effect).

5. Cytogenetic composition in the form of possession of blocks of heterochromatin. It is thought that the late replication of

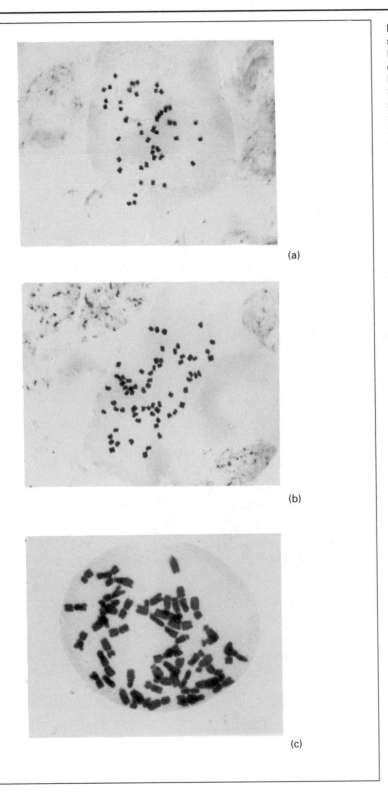

(a)

(b)

(c)

Figure 10.10 Chromosomal somaclonal variants. (a) Normal potato chromosome complement, $2n = 4x = 48$. (b) Potato regenerant, $2n = 67$. (c) Spread from a wheat cell suspension culture showing structural and numerical variation. Normal wheat $= 2n = 6x = 42$. Courtesy of Dr A. Karp.

heterochromatin affects the synchrony between chromosome replication at cell division, resulting in chromosome aberrations.

Somaclonal variation is frequently associated with chromosome instability such as structural and numerical changes (Figure 10.10). In fact, the ability of a plant to regenerate is directly related to the amount of chromosome variation present. Examination of callous chromosomes shows there to be more variation within the callus than the final progeny, implying that many of the chromosome changes are too severe to allow successful regeneration.

Taking into consideration the factors which have been shown to influence the level of somaclonal variation, several reasons (summarized in Table 10.2) have been proposed for the production of this phenomenon. Further study and understanding of these processes may enable more accurate control in the future.

Many examples of somaclonal variants exist, but those currently commercially exploited include new crunchier varieties of celery and carrot, now marketed as Vegisnax. Stable tomato varieties have been developed using somaclonal variants, one of which, DNAP17, was produced from commercial tomato variety UC82B and is resistant to *Fusarium* wilt owing to a single dominant gene mutation; DNAP7 from the same origin has a higher soluble solids content. Bell Sweet, a yellow bell pepper variety is almost seedless (an average of nine seeds per pepper compared to 330 seeds for a control variety, Yolo Wonder).

Somaclonal variation may also enhance the production of disease-resistant varieties. Passage of the regenerants through several tissue culture cycles in the presence of toxins has produced disease resistance. Examples of diseases to which crops are now resistant include:

- eyespot resistance:
  Helminthosporium sacchari, sugar cane;
  Helminthosporium sativum, wheat and barley;
  Helminthosporium victoriae, oats;

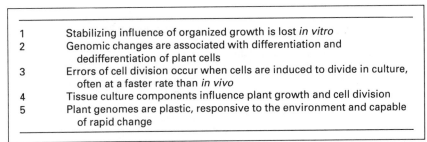

1	Stabilizing influence of organized growth is lost *in vitro*
2	Genomic changes are associated with differentiation and dedifferentiation of plant cells
3	Errors of cell division occur when cells are induced to divide in culture, often at a faster rate than *in vivo*
4	Tissue culture components influence plant growth and cell division
5	Plant genomes are plastic, responsive to the environment and capable of rapid change

Table 10.2 Reasons for somaclonal variation. Data taken with permission from Karp (1991)

- crown rust resistance: *Puccinia coronata*, oats
- Wilt disease resistance: *Fusarium oxysporum*, celery and alfalfa.

Herbicide-resistant varieties have also been isolated by this selective method.

In addition to production of improved agronomic characters, potential uses of callous culture include production of polyploid and aneuploid lines for breeding purposes. Polyploid lines have been demonstrated in tobacco, sugarbeet and potato. Aneuploidy has been exploited for production of new substitution lines in wheat alien lines. Induced translocations can be exploited to obtain introgression in hybrids, e.g. triticale and cotton. Biochemical mutants have been isolated from random mutational events and are useful for gene expression and development studies.

Somaclonal variation has made no significant impact on the breeding of major seed crops, such as wheat and maize. It is also unlikely to unless:

1. it can be more controlled;
2. it is proved that more variants of a particular type can be recovered using somaclonal variation than with conventional breeding methods.

This technique is currently advocated for crops whose breeding systems are limited and for less well-developed crops that have not been subjected to intensive breeding. This may be a suitable way to release new sources of variation, using a minimum of expensive technology. Hence, its use is currently being explored in many tropical and horticultural crops. The advantages and disadvantages of this technique are summarized in Table 10.3.

Table 10.3 Advantages and disadvantages of somaclonal variation. Data taken with permission from Karp (1991)

Advantages	Disadvantages
Novel variants may arise	Most variants are useless
Useful agronomic traits may arise	Not all changes are stable
Changes can occur in homozygous form	Most genetic changes segregate
Can be combined with cell selection procedures, for example herbicides and pathotoxins	Very few selection systems
	Characters of interest may not change and are virtually impossible to target
	Variation is cultivar dependent
	Non-reproducible

Box 10.9 Single-gene transfer techniques

So what are some of the more common methods for introducing single genes into organisms?

1. Microinjection. The isolated gene is present in linear form and is injected straight into the egg (just after fertilization). In mice, the nucleus is large enough to be directly injected. Fish zygote nuclei are covered by a tough egg coat (chorion). Injection is usually done through this, although in some species it is possible to digest it away. Because the nucleus is so small, injection takes place into the cytoplasm, not directly into the nucleus. With plants protoplast nuclei are microinjected. Direct injection of sequences with and without vector has been attempted in rodent muscle.

2. Electroporation. This can be used on both plant protoplasts and animal cells. When exposed to suitable external electric fields the cytoplasmic membrane starts to break down, i.e. 'pores' develop (hence the name). Conditions can be created such that the pores can reseal, but if DNA is in the surrounding fluid it can enter while the membrane is broken. Once the membrane is resealed, the DNA in the cytoplasm can enter the nucleus.

3. Viral vectors. A number of modified viral vector systems have been tested for use in human gene therapy. Retroviral vectors have been to the fore. They can transform actively dividing cells in culture with extremely high efficiency and stably integrate into the host genome without causing major rearrangements. However, viral vectors can only infect dividing cells, and concern has been expressed over safety aspects of this vector system, such as their potential to activate oncogenes. Alternative viral vectors are being developed to target slowly dividing (e.g. lung) and non-dividing (e.g. CNS) cells. The search for effective viral vectors continues, and it may well be that different types of viruses will have to be used in order to target specifically different areas of the body.

4. Microprojectile gun. Cornell University has developed a 'gun' that uses an explosive charge to fire DNA-coated tungsten pellets (1–4 μm in diameter) into embedded tissues or protoplasts. The DNA, once in the cells, has the chance of being integrated into the nucleus. This is a relatively new technique but gaining in popularity owing its wide-ranging non-species-specific application. Although initially designed for use with plants, it is now being adapted for mammalian use.

5. *Agrobacterium.* This is a vector-based system for use with plants. *Agrobacterium tumefaciens* and *A. rhizogenes* are soil-borne pathogens of plants that enter at the site of a wound and cause cell proliferation, resulting in a gall or tumour (*A. tumefaciens*) or hairy roots (*A. rhizogenes*). These are caused by a Ti (tumour inducing) plasmid contained within the *Agrobacterium* bacterial cell. On contact with the wounded plant cell, a segment of DNA from the *Agrobacterium* plasmid is transferred from the bacteria to the plant chromosome. Variants on these natural forms have been engineered for routine transformation in the laboratory and contain selectable markers, such as antibiotic resistance, to aid selection of transformed plants. This method has limited applicability as not all plants are susceptible to *Agrobacterium* infection.

6. Liposome-mediated gene transfer. Liposomes are fatty globules which fuse with cell membranes and so insert any DNA which is attached to them into the cell. It is currently under investigation as a means of delivering target genes for gene therapy. Liposomes containing the cystic fibrosis gene (Chapter 7) have been administered to patients in the form of a nasal spray, with some success in clinical trials.

7. Mammalian artificial chromosomes

Method	Organism	Organ/Organelle
Microinjection	Mammals other than humans	Nuclei
	Fish	Cytoplasm
	Plants	Nuclei
	Rodent	Muscle tissue
Electroporation	Plants	Protoplasts
	Mammalian tissue culture	Isolated cells
Viral vectors	Human	Specific tissues
Microprojectile gun	Plants	Tissue sections or protoplasts
Agrobacterium	Plants	Tissue sections
Liposome-mediated	Human	Specific tissues
MACs	Human	Under development

Table 10.4 Methods of gene transformation

(MACs). The theory behind MACs is that they are self-sufficient units that will integrate into a cell, replicating and dividing along with the normal complement. Because they do not have to integrate into the existing genome, mutagenicity fears are reduced. They are the mammalian equivalent of YACs (Chapter 2) and are in the process of being developed.

These methods are summarized in Table 10.4

10.4   Single-gene transfer (Box 10.9)

Single-gene transfer is the most intricate level of genome manipulation. This has only recently been feasible as the rapid development of molecular cloning techniques has enabled researchers to isolate specific gene sequences. Single-gene transfer is potentially the most wide-reaching of all the areas of genome manipulation as it can be applied to any organism and in a very specific manner. It is the only technique in this chapter that can be used in humans.

The idea of genetic transformation seems a simple one. It has the advantage that only one gene is inserted or altered, in contrast to mass transfer of genetic material in the form of whole chromosomes in conventional breeding. Manipulation of DNA in a test tube may be relatively straightforward. However, it must be remembered that each gene consisting of naked DNA has to be introduced into a complex cellular environment, find its way into the three-dimensional structure of the chromosome, integrate, express and still work appropriately. Also, if a permanent heritable change is required, germline transmission would have to be effected. The ideal situation is the integration of one complete gene which is expressed correctly. Needless to say, this rarely happens with any great level of efficiency. The other potential fates of foreign transformed DNA are summarized in Table 10.5 and described in more detail below.

Table 10.5   Possible fates of transformed DNA sequence

Integration of one complete copy
Partial or total degradation before integration
Transient expression
Non-release from vector transport systems
Integration after several rounds of cell division
Integration of multiple copies
Integration without expression
Adequate expression
Inappropriate expression
Genetic damage due to integration

- *Total of partial degradation before integration.* If the gene is injected into the cytoplasm and not the nucleus, the cytoplasmic exo- and endo-nucleases could digest it before it entered the nucleus. This is much more of a risk in those systems in which the DNA sequence is added alone without a vector. Partial degradation then directly affects the gene and is not offset by having a buffer of vector DNA.

- *Transient expression.* Some vector systems introduce the DNA into the cell as a circular molecule. This is capable of autonomous replication, but will not necessarily be integrated into the host genome. If integration does not occur it may be gradually lost during the processes of cell division and death as it cannot take an active role in the normal processes of cell division. Linear DNA can also linger in the cell for several rounds of division before being lost.

- *Non-release from vector transport systems.* The phagocytic and liposome transfer systems engulf the introduced DNA and surround it with a membrane. They then transfer it into the cell but may not always release it into the cell interior.

- *Integration after several rounds of cell division.* If the gene is transformed at the single-cell stage (microinjection into nucleus), integration after several rounds of cell division will produce a transgenic mosaic (chimera) with integration into some tissues and not others. The gene may not be integrated into the germ line and so be unavailable to be passed on to the next generation. The random nature of methods such as the particle gun used on tissue segments means that the resulting callous cultures are all mosaics. It is hoped that the plant can be regenerated from one of the callous sections expressing the transformed gene. With transformed mice, the chimaeras have to be backcrossed to the original parental strain and the offspring checked for heterozygosity of the transformed gene. Heterozygotes can then be crossed to produce homozygotic lines (if viable).

- *Integration of multiple copies.* Although many copies of the gene are used in the transformation experiment, the efficiency of integration is low. However, it is possible that several copies of the same gene may be integrated. Dosage problems may arise if multiple copies are all functional. This is reduced to a certain extent using homologous recombination. This is where the gene is added with large amounts of known genomic flanking sequence, the theory being that this new sequence will hybridize with the chromosomal sequence, integrate into the target DNA via recombination processes in the flanking regions, excising the old (non-functional) gene and replacing it with the new. Homologous recombination, thus, targets a specific region of the genome.

- *Integration without expression.* This may have many causes: degradation, mutation of gene or controlling sequences during the transformation process, lack of a satisfactory promoter, inactivation through DNA methylation or insertion into or next to a region of heterochromatin and subject to position effects (Chapter 2).

- *Adequate expression.* Expression has to be at sufficient levels to compensate for any mutant gene activity or in the case of transformed animals and pharmaceutical production, at a sufficient level to make the system economically viable. This may be influenced to a certain degree by choosing the right promoter and targeting expression at a specific organ. Not all genes need to work at full capacity, for example, the adenosine deaminase protein, which is essential for the normal functioning of the immune system, can maintain a functioning immune system at only 10% of its normal activity.

- *Inappropriate expression.* Related to the problem of adequate expression, the transformed sequence may integrate in a region of DNA controlled by either a developmental or tissue-specific promoter, limiting its pattern of expression. Some genes may require tissue-specific expression and therefore special targeting and tissue-specific promoters; others may require whole-organism expression. For example, ideally the cystic fibrosis gene would be targeted for expression in the lungs and the muscular dystrophy gene targeted for muscle expression.

- *Genetic damage as a result of integration.* However many safeguards are included in the vector system, there is always the fear that the gene may integrate into a functionally important sequence (causing more problems than it solves), for example integration into an oncogene may cause activation and subsequently cancer or some of the disarmed viral vectors used, although presumed inert, may mutate, reactivate and prove to be oncogenic. Fortunately, most organisms contain substantial proportions of redundant sequences into which the gene would invariably integrate, if homologous recombination techniques are not used.

While the list of genes suitable for transformation experiments is increasing all the time, the number of successful examples is fairly limited.

Human gene therapy, although in its infancy, is no longer science fiction. The gene disorders targeted initially are single-gene defects. The first trial was approved in the USA in September 1990 to replace a mutant adenosine deaminase (ADA) gene with normal copies in patients with severe combined immunodeficiency disease (SCID). This was effected by removing a sample of the patient's T-lymphocytes, incubating this sample *in vitro* with the ADA gene in a retroviral vector, increasing the bulk of transformed cells in culture and then transferring them back to the patient. Initial results show a definite clinical improvement. Recently permission has been granted both in the UK and in the USA for trials of the cystic fibrosis gene therapy, introduced into the lungs via a nasal spray containing the gene in either an adenovirus or liposomes. This is seen as the best way of attacking the lungs, while minimizing potential problems with whole-body gene expression. Many more trials have been approved since, but these are all of a very limited nature, and no astounding successes have so far been achieved. Part of the problem with conducting clinical trials for gene therapy is in finding patients who are not already undergoing some treatment regime. The doctors have to be able to prove that any effect is due to the gene therapy and not caused by any previous treatment, so most patients are ruled out and hence sample sizes for gene therapy are very small.

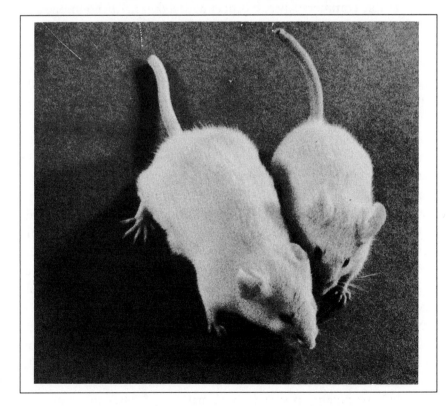

Figure 10.11 'Giant transgenic mouse' with its normal counterpart. This was produced by genomic integration of the metallothionein growth hormone. Reprinted with permission from *Nature*, **300**, No. 5893, cover.

Work on animals first came to the fore with the production of the 'giant mouse'. A mouse was microinjected with a growth hormone gene, which when activated produced a mouse double the size of the original strain (Figure 10.11).

Great interest lies with transgenic mice as model systems, for example in producing them as test models for drug therapy. Proteins with which drugs interact are introduced into transgenic mice and used to predict the activity of drugs in certain disease states. For example, cancer cells often develop multiple drug resistance. Only a single gene is responsible for this condition, the multidrug transporter or P-glycoprotein (MDR-1). Mice expressing the gene for this protein are multidrug resistant and used to test novel chemotherapeutic agents.

Mice are also used to mimic human diseases. The normal mouse gene is either totally inactivated or mutated via homologous recombination ('knock-out mice'). The resulting physiological effects are studied as a possible model for the human disease and to indicate potential therapies. The most famous example of this is the Harvard 'oncomouse', which readily expresses the oncogene *myc* and is very susceptible to tumours. It is available commercially as a cancer model (*myc* has been implicated in breast cancer). Knock-out mice now exist for many human disease conditions, such as cystic fibrosis, artherosclerosis, Duchenne muscular dystrophy, sickle cell anaemia and β-thalassaemia, and although they do not always mimic the human condition accurately they are generally considered the essential first stage in the genetic analysis of gene function.

Other animals of interest include milk producers. The aim here is to produce transgenics which excrete valuable chemicals in milk. The first transgenic sheep was produced in 1989. It secreted human anti-haemophilic factor IX, albeit in very small amounts. The most successful example so far is 'Tracey, the transgenic sheep'. Tracey excretes 1.5 kg of α_1-antitrypsin in her milk per annual lactation. α_1-Antitrypsin deficiency results in emphysema, a progressive lung degenerative disorder. Each patient requires 200 g per year and the current cost is prohibitive. So Tracey still has quite a long way to go before commercial production is viable. The ultimate aim is to produce transgenic cows, as these produce 10 000 of milk per year compared with 250–800 l from sheep or goats. Unfortunately, transformation of cows involves cumbersome and costly surgical procedures, so sheep and goats are the only systems available at the moment.

Early experiments to speed up animal growth rates using growth hormone genes have so far met with failure. Transgenic pigs suffered from a range of orthopaedic disorders as a result of the increased growth rate. Similar experiments on fish have had more success. Initially the fish were transformed with rodent growth hormone genes, as these were the only ones available. Subsequently the piscine equivalents have been cloned and used. These function more effectively as the control sequences are more easily recognized and are also, from the point of view of the general public, more acceptable. Transgenics is a controversial issue, but more so if genes used are cross-species.

The first field trial of a transgenic plant was in 1986 in the USA. There have been 860 field trial approvals since, in Europe and the USA, but only now are products coming to the commercial market. The USA will soon see the launch of Flavr Savr™ tomato (containing the antisense gene for ripening) and virus-resistant squash. It is expected that cotton transformed with the *Bt* (*Bacillus thuringensis*) gene making it resistant to attack by lepidopteran insects will also soon be launched. The main characters transformed include herbicide, viral and insect resistance, quality of fruit, male sterility and stress resistance.

While the potential for plants is probably greater than that for humans and animals, they are subject to far more consumer mistrust and resistance. The hardest challenge may well be getting transformed products accepted by the general public, rather than the technical procedures themselves. The Flavr Savr™ tomato is initially going to be used in tomato-based food products as the general public is thought to be fairly hostile to buying genetically engineered food. This has led to a huge debate, particularly in the USA, about whether food containing genetically engineered organisms should be labelled in the same manner as food additives.

Much debate currently centres around the ethics of genetic engineering, of both plants and animals. In response to public concern, new government committees have been set up specifically to deal with the issues surrounding gene therapy. It can take up to 2 years to gain approval for gene therapy studies and equally as long if modifications to the approved procedure are required. The granting of patents on genetically engineered organisms is particularly controversial. The first patent granted on an animal by the USA was Harvard's 'oncomouse' in 1988. This decision was upheld by the European patent court 3 years later. Since then, three more patents have been granted for transgenic mice, with over 180 more in the pipeline. Hence, a major trend among lawyers is developing an interest in genetic engineering! What is certain is that these techniques allow for the most sophisticated manipulation of DNA and are here to stay. Having said that there are many situations where single-gene transfer is not appropriate or feasible (e.g. polygenic inheritance) and more conventional methods may be used.

Historical aspects of cytogenetics

11

Summary

Science is not, and never has been, an activity carried out by an intellectual elite for the interest and benefit of the ruling classes. The very best science has always enlightened both the individual and the society that supports it. The actions of science have always been, in the past, associated with enlightenment. It is important that those who study science, and the fact that you are reading this implies that you are a scientist, take responsibility for their actions. You should be aware of the past successes and failures and be determined that misunderstandings will not fog the image of your work. If you do not communicate what you know in an intelligible manner, then you may as well not know it in the first place; you will have wasted your resources.

History is not just for historians; we can all learn from the past. The hubris of Henry Ford's quote 'history is bunk' stands as a monument to stupidity. Ford worked on machinery which was invented not out of thin air, but from decades of research by other people. Such work had to be published to put it into the common domain. If we all started from basic principles we would be no more advanced now than when intelligence first differentiated man as a thinking being.

Every time you read a bibliography at the end of a scientific paper you are looking at a brief and idiosyncratic history of the subject. Historical interpretation of science is vital for the correct attributions of events. We are all indebted to the past for work which we do not need to repeat.

There are many strands to the history of cytogenetics: microscopists, medics, physiologists and botanists all contributed to the foundations of modern cytogenetics. At the time many of these contributions appeared to be unrelated. It is only with hindsight that a coherent overview of the key events in the developing field of genetics is possible.

Before the advent of cytogenetics and the study of chromosomes, a concept of cells and a theory as to the relationship between cells and bodies, cell theory, had to be developed. It should always be remembered that the information which we take for granted has frequently been hard won. Without adequate techniques of microscopy (Box 11.1) it was quite impossible for anyone to know that multicellular organisms were just that and not a large-scale structure made up of different but homogeneous units.

Theodor Schwann and Matthius Jakob Schleiden are credited with the original definition of cell theory in 1830. The two scientists were friends, although working on completely different material. Matthius Jakob Schleiden was a botanist working in Hamburg. Using compound microscopes and the relatively primitive preparative techniques available at the time, he recognized the presence of cells in plant tissues and made accurate deductions about both their importance and universal nature.

Theodor Schwann was a remarkable scientist, at that time training in physiology. He described what became known as the Schwann cell, a myelinated nerve cell. As well as this he made a pivotal discovery in embryology with his description of the single-celled nature of the egg. He demonstrated that spontaneous generation of life was impossible in sterile broth, although this was not completely accepted until the early 1860s, when the work of Louis Pasteur echoed Schwann's findings. Unfortunately, attacks on his reputation by influential Germans caused him to emigrate to Belgium, where he became a mystic and largely gave up science.

As described by Schleiden and Schwann, cell theory contained one important mistake: it contended that new cells arose by budding from an existing nucleus, not by cell division as it is now understood.

Science and technology developed rapidly in the middle of the nineteenth century. Key defining points in the history of biology include the publication in 1858 of Charles Darwin's and Alfred Russel Wallace's papers describing ideas that were encapsulated a year later in *The Origin of the Species*.

Box 11.1 Microscopes and chromosomes

While it is commonplace to think of cytogenetics as a new science, it should be remembered that it was in the nineteenth century that the first developments in the investigation of chromosomes took place. The work resulted from the development of microscopes and preparative techniques, but the groundwork had been laid even earlier.

Optical microscopes attained what we would normally regard as modern standards of resolution in the middle of the nineteenth century. This accounts for the discovery by Flemming of lampbrush chromosomes in 1882 and the discovery by Balbiani of polytene chromosomes in 1881.

To appreciate the great implications of the use of microscopes it must be remembered that chromosomes are, on a human scale, ex-tremely small. Consequently, without microscopes, little would have been discovered about chromosomes.

One of the very first microscopists, Antony van Leeuwenhoek (pronounced lay-ven-hook) was Dutch and manufactured his own lenses. He was born in 1632 and lived until 1723, an extremely long time then.

Compound microscopes were in use before 1650 but because of poor

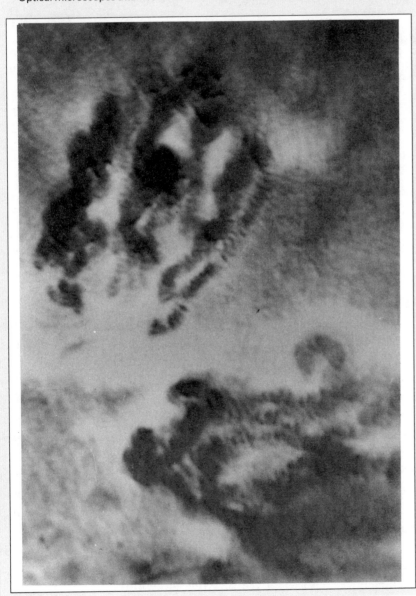

Figure 11.1 Banded chromosomes and the simple microscope. Techniques old and new meet in an intriguing experiment by Brian J. Ford. Chironomid giant chromosomes are imaged through a simple (= single lens) microscope. The micrograph demonstrates how much the pioneers of cytology could have seen, using instrumentation available since the seventeenth century. The squash preparation of salivary gland is stained with acetic orcein and mounted in DPX, and the microscope (similar to the type perfected by Leeuwenhoek in the 1670s) contains a single magnifying element rated at 295 ×. Micrograph courtesy of Brian J. Ford.

lenses were of relatively little use. Leeuwenhoek chose to develop single-lens microscopes, which he produced to a very high standard by polishing small glass droplets.

One of the reasons that simple microscopes were not highly thought of is summarized by the memorial note written by Martin Folkes, Vice President of the Royal Society, on the death of Leeuwenhoek. In this article Folkes implored Fellows of the Society not to distrust the conclusions of Leeuwenhoek just because they could not confirm his results.

This very astute request by Folkes reflected the remarkable powers of observation which Leeuwenhoek possessed. The photograph of polytene chromosomes (Figure 11.1) was taken using a single-lens microscope. Although not of modern resolution, we can imagine that it would have been possible for Leeuwenhoek to have seen chromosomes had the preparative techniques been available.

Little information is available regarding the particular equipment used in the early studies of chromosomes, but in 1879 Zeiss introduced the oil emersion lens designed by Ernst Abbe, a German physicist. This allowed far more detail to be seen in the cell interior.

In 1884 E.M. Nelson had attained the ultimate resolution of the standard light microscope using a no. 1 stand of Powell and Lealand. Unfortunately, such was the commercial pressure from the Zeiss no. 1 'jug-handle' microscope that by 1910 Thomas Powell had virtually given up the manufacture of microscopes.

It was essential that improvements in microscopy were paralleled by improvements in preparative techniques; no lens can yield information from badly prepared specimens. One such major step forward was the introduction of the automatic advance microtome. Automatic microtomes for cutting serial sections had been used since about 1770, but in 1885 Cambridge Scientific Instruments introduced the robust and relatively cheap Cambridge rocking microtome. This produced ribbons of wax-embedded serial sections of even and reliable quality.

In 1865 Gregor Mendel delivered several papers at the Brunn Natural History Society. The transcripts of these were published in 1866.

Mendel had a rather unlikely start to his career. He was the son of a peasant farmer and joined the Augustinian monastery in 1848 at the age of 21. He tried twice to pass the teaching examinations of the order, but failed both times. In 1868 he was elected Abbot and thereafter virtually gave up the study of plant breeding which had resulted in what we now refer to as Mendel's laws. It is unfortunate that it was many years after his death in 1884 before the importance of his results were appreciated (Box 11.2).

Early in the nineteenth century Ronald Fisher played a central role in reconciling Darwinian theory and Mendel's results. He formulated the theoretical basis of the statistical analyses of genetics. When Fisher looked at the results presented by Mendel he found them statistically too precise.

This is not surprising. Mendel had looked at over 21 000 plants in his experiments and had a detailed knowledge of his experimental organism. He was using his results to demonstrate the mode of inheritance of specific characters, which he fully understood. Those results that were not interpretable or which deviated from his expectations would have been left out. Mendel's knowledge of his organism can be assumed because from the number and range of characters which he dealt with it is almost certain that he would have come across the phenomenon of genetic linkage. Presumably, because this had no explanation at the time, Mendel never reported any linkage phenomenon.

1840 and all that
The year 1840 saw the introduction of a uniform penny post in the UK and the colonization of New Zealand. Pillar-boxes for posting letters in Great Britain were introduced by a senior civil servant of an idiosyncratic turn of mind. When he left the civil service in 1867 Anthony Trollope developed the writing skills for which he is so much admired today. In 1843 a steam ship, *Great Britain*, crossed the Atlantic using a propeller for propulsion, and 5 years later in 1848 Marx and Engels published the Communist Manifesto.

Box 11.2   Mendelian inheritance

Although Mendel did much of his work on sweet peas, the principles are the same for any monohybrid cross. It should be remembered that this is not always the case as there are many situations of incomplete dominance, penetrance and polygenic inheritance.

The nomenclature is simple: capital letter for the dominant trait, lower-case letter for the recessive trait. For example, in the case of shallow-leaved and deep-lobed leaves of *Coleus*, deep is dominant and so is designated D, shallow is recessive and so is designated d. This apparent lapse in logic does in fact have a reason; it reflects the homology between two apparently different phenotypes.

In a cross between pure-bred deep and pure-bred shallow the following situation is observed.

```
Deep-lobed leaves (DD)        ×        Shallow-lobed leaves (dd)
Gametes all D                          all d
F1                  all heterozygous deep lobed
                              Dd
Crossing F1 × F1
             Deep-lobed                 Deep-lobed
                Dd                          Dd
Gametes D and d                    D and d
F2                  DD Dd dD dd
             Three deep-lobed : one shallow-lobed
```

As can be seen, the F2 generation has a phenotype of three deep-lobed to one shallow-lobed, but this does not reflect the genotype of one homozygote for deep lobes, two heterozygotes and one homozygote for shallow lobes. It does demonstrate the difficulty in determining whether an individual is homozygous or heterozygous for a dominant trait.

Although this functionally explains Mendel's observations it does not explain *why* some genotypes are dominant and others recessive; that is a different and very much more complicated problem.

Also in the second half of the nineteenth century, chemistry was being applied to biology. In 1869 a Swiss chemist, Friedrich Miescher, analysed cell extracts and demonstrated that they were made up of protein and a nucleic acid; the science of this is dealt with in more detail in Chapter 1. It was soon after this that Albrecht Kossel analysed nucleic acid and identified the basic components. It was he who named the five bases adenine, guanine, cytosine, thymine and uracil.

The great cell biologist Walther Flemming (1843–1905) is reputedly the first person to have seen chromosomes, although he did not coin the word. As a very skilled observer, using the new techniques of histochemistry and the high-quality optics available from several manufacturers, Flemming was able to develop a clear idea of the processes that occur during cell division. Until then there had been a great deal of confusion surrounding these processes. He was most certainly aware of the, as yet, unnamed chromosomes since he based much of his descriptive work on nuclear division on the activities of chromosomes. It was Flemming who, in 1882, coined the term mitosis. He accurately perceived that after cell division each daughter nucleus contained as much chromatin as the original cell. Flemming serves well to illustrate the importance of choosing an experimental organism that satisfies all the necessary criteria. Flemming used what may at first sight seem an odd choice; epidermis of salamander embryos. This is an inspired choice for several reasons, two of which serve to demonstrate this. The first is that although the embryo is small the cells that make it up are large. The second is that, difficult though it is, it is possible to see metaphase taking place in unstained and therefore live material. As can be appreciated, sorting out the various stages of cell division from the 'snapshot' afforded by cells fixed at different stages of mitosis was no easy task.

The term 'chromosome' was coined by Wilhelm Waldeyer-Hartz (1839–1921) in 1888 to describe the 'coloured bodies' of the nucleus.

The stages of nuclear division that we now describe as prophase, metaphase and anaphase were defined by Eduard Strasburger, a botanist of immense ability. He worked on the hair cells of *Tradescantia*, a common and easily cultivated house plant.

Once chromosomes had been identified one of the earliest questions to be asked was 'how many chromosomes are present in cells and is the number really constant?'. This question was asked particularly about human tissue. In 1891 Hansemann was working with human material and said that he had seen cells with 18, 24 and more than 40 chromosomes. There were two main reasons for this range of numbers. The first was that material was usually sectioned so that interpretation of images was very difficult indeed. The second was that it was not automatically assumed that the number was constant. It was also in 1891 that Henking made the striking observation that in some insect sperm there was an additional nuclear structure occasionally to be found. He referred to this as the X-body to reflect the mystery of the object. In the following years estimates of human chromosome numbers varied from 16 in 1892 to 32 in 1906.

Although not initially recognized as of great significance to genetics, but instantly recognized as important to the applied science of medicine, the work of Karl Landsteiner in 1900 was seminal. Until this time the outcome of blood transfusions was widely variable, sometimes complete cure and sometimes death. The only method of control was to use donor as closely related to the recipient as possible. Landsteiner demonstrated the agglutination that could result from inappropriate mixtures of blood. He reasoned that this was probably what happened in the body and could be explained on the basis of presence or absence of cell-surface antigens associated with the red blood cells. The two antigens were termed A and B. Either of these could be present, giving either blood group A or B; both

---

Box 11.3 Alfred Bernhard Nobel

Alfred Bernhard Nobel was a Swedish chemist born in Stockholm in 1833. In 1850 he went to Paris to study chemistry and travelled widely before going to the USA to work with another Swede, John Ericsson. In 1829 Ericsson built a rival to Stephenson's Rocket (built the same year) and in 1836, 6 weeks after Sir F.P. Smith, he patented one of the first successful screw-propellers.

After returning to Russia where he was brought up, Nobel settled in Sweden in 1859. He was, like his father, interested in the civil engineering uses of explosives, so in 1865 he started manufacturing nitroglycerine. Unfortunately, that same year the factory blew up, killing five people, one of whom was his brother, Emil. The major step forward in the fortunes of Nobel was the discovery in 1866 that nitroglycerine can be handled in relative safety if it absorbed onto kieselguhr, a diatomaceous earth that was used for polishing and chromatography. This was sold in waxed card tubes under the trade name Dynamite, reputedly the most widely known trademark during the first half of the twentieth century. A further development was the use of nitrocellulose as the absorbent material, sold as gelignite.

Nobel built up a massive manufacturing empire based on more than 350 patents ranging from artificial gutta-percha, to armour plating. When he died in 1896 his fortune went to set up the Nobel Prizes, the first of which was awarded in 1901. Originally there were five categories, physics, chemistry, medicine or physiology, literature and peace. A sixth prize, for economics, was instigated in 1969.

The synthetic transuranic element 102, nobelium, is named in his honour.

---

could be present, giving group AB, or neither, giving group O. Using this idea a simple test that reduced the risk of blood transfusions was devised. It quickly became obvious that inheritance of blood groups followed a very straightforward mechanism. Landsteiner was awarded the Nobel Prize in 1930 for this invaluable work (Box 11.3).

It should be remembered that heredity as a subject for legitimate speculation had been around for a long time! The real problem was the lack of any usable rules governing inheritance. Everyone could recognize a relationship between parents and offspring; seeds from one crop plant always gave rise to the same type of plant, just as domestic animals always gave rise to their own type. This was the very reason that confusion reigned, as it was also noted that there were subtle differences between generations; generations are similar but not identical. As a result it became axiomatic that inheritance was based on the untestable principle of blended characters.

Genetics received a considerable boost in other ways in 1900. Three scientists were independently pondering the process of inheritance: Hugo Marie deVries, Karl Franz Joseph Correns and Erich Tschermak von Seysenegg. As is normal under these circumstances, they each undertook a literature search to see if anyone else had been studying the same topic. They were understandably surprised when each of them came across the work of Mendel from 40 years earlier.

DeVries, the son of a Dutch prime minister, studied medicine originally, but later changed to botany and conducted many breeding experiments in this area. It was quite clear from his results that he had a 3:1 ratio appearing with some of the characters which he was studying. He also came across ploidy changes in *Oenothera*, the evening primrose, although he mistakenly thought that it was very fast mutations that were causing the observed changes in phenotype.

Correns, also a botanist, was born in Munich. He, too, came across the ratios of heredity by chance.

Although these three were the rediscoverers of Mendel's work, it was William Bateson who most vigorously pursued the ideas of Mendel. There was opposition to the ideas contained in the work simply because of the newness of the subject and the fact that such interpretations which were made were not necessarily self-evident at that time. Indeed, these ideas were often seen as counterintuitive. Bateson had come to the conclusion that the process of heredity was one not of continuous change but of discrete jumps. These ideas were published in 1894 in a treatise entitled *Materials for the Study of Variation, Treated with Especial Regard to Discontinuity in the Origin of Species*. When he discovered Mendel's work he realized that it deserved a wider audience and so he arranged for the publication of a translation in *The Journal of The Royal Horticultural Society*, to which he added annotations. Although Bateson coined the word 'genetics', he found it difficult to accept the idea of natural selection or the central position of chromosomes in heredity.

It was in 1903 that chromosomes first took centre stage in genetics. Walter Sutton, an American, and Theodor Heinrich Boveri, a German, had proposed in 1902 that chromosomes were important, and in 1903 Sutton published *The Chromosome Theory of Heredity*. Boveri originally worked on *Ascaris* eggs, but later turned to sea urchin eggs as an investigative subject. His choice of organism was fortuitous and he conducted a series of experiments that rival the very best in manipulative techniques. Boveri, using sea urchins, demonstrated that a full complement of chromosomes were needed for development of the embryo. He did this by surrounding the eggs with a mass of sperm, a situation not normally found in the wild. This resulted in occasional fertilization of the egg with two sperm. As can be readily appreciated, the result was disrupted mitosis with some daughter cells having too many chromosomes and some having too few. Boveri showed that embryos with an abnormal chromosome complement developed abnormally.

Sutton was also working on material which at first glance would not appear to be suitable for chromosome research; grasshopper testis. Grasshoppers, as used by Sutton, have a diploid number of 23. These

1903 and all that
In 1903 the Wright brothers made their first manned flight at Kitty Hawk in an aeroplane called *Flyer*. Their longest flight lasted just 59 secs. This should not be taken as an achievement without precedent: in 1890 Clement Adler first demonstrated a full-sized aircraft that could take off under its own power. In 1896 Langley flew a steam-driven machine for over a kilometre, after which it crashed. It was in 1901 that Whitehead achieved the first flight on an aeroplane which was motor powered. Such is the level of human ingenuity that by 1976 commercial flights of Concorde had started, carrying passengers faster than the speed of sound.

> **Box 11.4 Dyes, microscopy and innovation**
>
> Although microscopes had been available for many years, their use was limited by the techniques of preparation that were available. It should be remembered that microscopes have three properties: magnification, resolution and contrast. Surprisingly, of these three the least important is magnification. This is not counterintuitive, as might at first be thought. Without both contrast and resolution there is little or nothing that can be seen, regardless of magnification.
>
> To generate the essential contrast and consequent resolution of detail, dyes are normally used. It is, of course, possible to visualize details of internal cellular structure by using sophisticated techniques of phase contrast. This was not, however, possible in the early years of microscopy.
>
> When Leeuwenhoek was producing his single-element microscopes Hooke had already produced low-resolution compound devices. Realization of their potential as instruments of investigation was confined to areas where natural contrast was highest. Leeuwenhoek was the first person to use a stain in microscopy. This was saffron extracted from the autumn crocus and was used to stain muscle cells. Other dyes that were valuable in histology were natural products such as cochineal and indigo.
>
> This use of natural dyes continued until the synthesis of the first artificial dye in 1856. This major step forward was made by William Henry Perkin (later Sir) when he was 18 years old. He had been trying to synthesize quinine by oxidizing aniline, which was not possible, but left a coloured residue characterized as mauve. The resultant mauve dye was a great success. By building his own factory to manufacture the dye he was in a position to retire at 36 and devote himself to research.
>
> From this time new synthetic dyes were being made all the time. Basic fuchsin was produced in 1856, safranin in 1859, methyl violet, aniline blue and spirit blue in 1861. In 1871 one of the most widely used stains, eosin, was produced while methyl green was first synthesized in 1876.
>
> It should be appreciated that none of these dyes was manufactured specifically for use in histology. The main use of these new dyes was in the textile industry. Perkins' mauve was, for example, primarily used to dye silk. Because of this principal use dyes were made to an exact colour, but not an exact chemistry. In 1922 the first systematic nomenclature of dyes was introduced. This was produced by the Society of Dyers and Colourists in Bradford and was based on chemical structure. A single colour index number distinguishes the dye; thus, basic fuchsin is 42510, methyl violet 42535 and methyl green 42590.

are made up of 22 autosomes, which Sutton demonstrated could be made up into 11 distinct pairs, and a single accessory chromosome. C.E. McClung had suggested that this single chromosome may, in some way, be involved in sex determination. The accessory chromosome was easily discernible from the remainder of the chromosome complement and helped Sutton to the conclusion that there were two types of sperm produced and that there was a requirement for a complete set of chromosomes (11) from each parent. Chromosomes had finally been given a purpose which could be experimentally tested.

Bateson demonstrated in 1905 that some characters do not segregate independently. It was also the point at which Clarence McClung was able to show that in mammals females have two X chromosomes whereas males have a single X and Y.

Thomas Hunt Morgan began his pioneering work on *Drosophila* in

> **Strowger Telephone Exchange**
>
> In 1905 Almon Brown Strowger, an American undertaker, convinced that switchboard operators were rerouting his custom to rival companies, invented the direct dial telephone. For the next 80 years the Strowger mechanical exchange would control the flow of telephone calls until the digital electronic exchange was introduced.

1907. This would establish both the chromosomal basis of inheritance and the nature of the linkage that Bateson had previously described. It also set a precedent for using *Drosophila* as an experimental organism that extended beyond genetics and into embryology.

In 1908 attitudes towards genetic disease were shaken up. Sir Archibald Edward Garrod published a paper describing inborn errors of metabolism. While studying alcaptonuria, albinism, cystinuria and pentosuria Garrod showed that there was incomplete metabolism of a dietary component, the part product being unexpectedly excreted in the urine. By studying affected families he showed the first examples of mendelian recessive inheritance in man. Taking one step further he suggested that the problem was a defective gene affecting a metabolic pathway. It would be 50 years before this theory was shown to be correct, the disease being due to a defective enzyme.

In this scientific revolution the old vocabulary of biology was no longer adequate for the task of dealing with genetics. New terms were being coined all the time, such as 'gene' and 'chiasmata'. It was also in the year 1909 that Soren Sorensen defined a scale which has been of immense practical value to scientists: pH.

In 1910 T.H. Morgan, whose grandfather F.S. Key wrote the American national anthem, published data that introduced the concept of sex linkage using red-eyed and white-eyed *Drosophila*. He had not originally been completely convinced by the work of Mendel, but his own work with *Drosophila* soon convinced him of its truth. His extensive work won him the Nobel Prize in 1933, but before then, in 1911, he had produced the first ever linkage map of *Drosophila* showing five sex-linked genes. By 1922 this map had been extended to show over 2000 genes.

In 1912 estimates very close to the true number of human chromosomes were made by Von Winiwarter. He was studying spermatogenesis and had seen metaphase plates with 47 chromosomes made up of 23 pairs with a single unpaired chromosome left over. His conclusion was that males had 47 chromosomes and females 48. Since it was already known from the work of McClung that some insects had an XX–XO sex-determining mechanism, this result was not unacceptable. Since we do not know the origin of his material it is perhaps possible that he had obtained abnormal, aneuploid, material. In the same year H.L. Wienmann published a paper in which he concluded that there were 34 chromosomes in humans and, in contradiction to Von Winiwarter, that sex determination was not XX–XO.

In 1919 T.H. Morgan published *The Physical Basis of Heredity* and in 1920 Hans Winkler coined the word 'genome'.

Box 11.5 Svedberg and the Svedberg unit

Theodor Svedberg (1884–1971) studied and worked all his life at Uppsala University. Svedberg developed one of the fundamental machines in cell biology – the ultracentrifuge. For this and the work to which he put it he received the Nobel Prize in 1926.

Svedberg's key insight was that, although it was commonplace to use a centrifuge to spin down large particles, all molecules have mass and are therefore susceptible to the force of gravity. Simply by increasing the gravitational force it should be possible to pull down molecules of progressively smaller molecular weights.

The first use of Svedberg's machines was to confirm a theory of Herman Staudinger (1881–1965), a chemist whose work on rubber developed into modern polymer chemistry. This earned him a Noble Prize in 1953. Staudinger thought that, contrary to contemporary ideas, materials such as rubber were made up not of disordered small molecules, but of large molecules joined together. Using this logic proteins could also be large polymers. Using Svedberg's ultracentrifuge such large molecules became separable.

Svedberg's ultracentrifuges sediment by the power of gravity (g). His machines ran at a speed of up to 140 000 r.p.m., which generated gravitational fields of 900 000 g. Because coefficients of sedimentation are very small they are now expressed in Svedberg units (S). $1S = 1 \times 10^{-13}$ s. Solvent, of course, affects sedimentation, especially if it is made up of a density gradient. On this basis it is normal to express S in terms of water at 20°C. In this way the unit S gives a comparative measure of both size and molecular weight.

1921 and all that

Czech playwright, Karel Capek, published a play in 1921 called R.U.R. which was to introduce a new word into the language. R.U.R. stands for Rossum's Universal Robots. Benito Mussolini gained power in Italy in 1922.

In 1923, Theodor Svedberg developed the ultra centrifuge (Box 11.5).

Work by Theophilus Painter, published in brief in 1921 and in more detail in 1923, gives us an insight into the methods of investigation available to workers using human material. Painter used material from three individuals who were being castrated in a Texas state institution for 'excessive self-abuse coupled with certain phases of insanity'. He found cells with between 46 and 48 chromosomes and, although he argued for an XX–XY mechanism, he did not lay to rest the question of sex determination in humans.

The standard work for geneticists on statistics was published in 1925 by R.A. Fisher. It is called *Statistical Methods for Research Workers*. By 1926 that prolific and accurate geneticist Morgan had published more groundbreaking work, *The Theory of The Gene*.

In 1927 Muller and Stadler described the mutagenic effects of X-rays on genes, for which Muller was awarded the Nobel Prize in 1946. A year later (1928) Fred Griffiths published a paper in the *Journal of Hygiene*. This described the transformation of smooth colonies of pneumococcus to the rough form by transmission of genetic material between cells. Griffiths was unfortunately killed in an air raid on London in 1941.

By 1940 biochemical genetics was well established and G.W. Beadle and E.L. Tatum decided to investigate the pink bread mould *Neurospora crassa*. This was a very useful organism for several reasons, such as ease

1927 and all that

In 1927 Lindbergh made the first non-stop flight across the Atlantic Ocean, taking 33 hrs and arriving in Paris to win the $25 000 Orteig Prize.

of cultivation and speed of growth. But *N. crassa* has one particular advantage: the adult stage is haploid. This, of course, means that all mutations are expressed. By irradiating various strains of this fungus its ability to metabolize different nutrients in specially developed growth medium could be investigated. It was this work which culminated in the 'one gene, one enzyme' hypothesis. Although this has since been considerably modified, it does still serve as a useful concept in most cases. Interestingly, this idea had been arrived at many years earlier by Garrod. Joshua Lederberg went to Yale University to work with Tatum on bacterial genetics in 1946, and in 1958 all three of these workers shared a Nobel Prize for the 'one gene, one enzyme' hypothesis.

By 1944 there were many workers who styled themselves 'geneticists'; and as a subject genetics was well established but there was still a massive gap in understanding of basic principles. What precisely constituted the genetic material? It was known that chromosomes carried genes and linkage maps had been around for about 30 years, so one might imagine that DNA was known to be the genetic material; this was not so. The controversy was considerable and difficult to resolve. Chromosomes were known to be made up of histone and non-histone protein and nucleic acid, but whether it was one or all of these that formed genes was completely unknown. Using straightforward reasoning it was easy to rationalize the idea that it was protein that carried genes. Chemically, protein is complex, nucleic acid is simple. Genes are complex, therefore genes must be protein. There was much debate over this, but in 1944 a paper appeared in the *Journal of Experimental Medicine* written by O.T. Avery, C.M. Macleod and M. McCarty detailing experiments that culminated in experimental evidence that genes were made up of DNA; a more detailed explanation of this is given in Chapter 1. The certain knowledge that DNA was the very stuff of heredity fuelled interest in genetics. It was now important, scientifically, to deduce the arrangement of the material and the manner in which DNA works.

From a rather more applied angle the first UK genetics clinic was opened at Great Ormond Street Hospitals for Children in 1946.

In 1947 Tatum and Lederberg discovered bacterial recombination. At the same time Barbara McClintock was working on the genetics of maize. The puzzling results that she generated gave rise to many novel ideas, for example that there are genes that specifically control other genes and that physical movement of genes (jumping genes or transposons) is quite normal in nature. For this work McClintock received the 1983 Nobel Prize for physiology or medicine. Transposons are described in Chapter 6.

When Murray Llewellyn Barr described a nuclear inclusion in 1949 which could be detected in female mammalian cells but not in males, it was natural to refer to it as a Barr body. Barr bodies turned out to be the condensed and inactivated second mammalian X chromosome. The importance of this process in sex determination is discussed in Chapter 8. Mary Lyon was later to describe the theoretical basis for this process of random inactivation of X chromosomes in female cells.

During the years after 1950, the base equivalence of DNA was being worked out. It was in 1953 that the now famous double helix structure of DNA was elucidated by J.D. Watson and F.H. Crick, for which they won the Nobel Prize in 1962 along with Maurice Wilkins. The structure of the double helix is described in Chapter 1. The often repeated controversy regarding the work of Rosalind Franklin not being similarly honoured is not in fact a controversy at all. Franklin died in 1958 at the sadly early age of 37; the Nobel Committee is not allowed to award Nobel Prizes posthumously.

With the ushering in of the second half of the twentieth century and the concomitant peace and stability of Europe, science became better funded. More important than this, though, was the realization that biology in general, and genetics in particular, are as technically demanding and stringent subjects as chemistry or physics. The next decades were going to be the province of the geneticist. Genetic manipulations were going to delineate areas of contention and debate as new techniques impinged on everyone's lives.

In 1956 Tjio and Levan laid to rest the debate as to the number of chromosomes in a normal human (modal number 46). This opened up possibilities for prenatal diagnosis of genetic disease. Tjio and Levan made a bold move when they published their results, having used the very modern techniques of treating cultured cells in various ways to make the chromosomes more readily visible. Although Kemp had cultured embryonic material in 1929, treatment of cultured cells with spindle poisons and hypotonic solutions was a long way off.

Tjio and Levan recognized 46 as the important number, even though 48 was by then assumed to be correct. Later it was found that another group at Lund, Hansen-Melander, Melander and Kullander, had been working on human liver in tissue culture, but being unable to find 48 chromosomes they abandoned the project; the number they repeatedly found was 46. The first successful prenatal diagnosis was carried out in 1966, although biochemical screening for alkaptonuria was started in 1961. The first human gene, human placental lactogen, was cloned in 1977, and a year later the first prenatal DNA diagnostic test was introduced for sickle cell disease. By 1988 the techniques that had been developed were so sophisticated that it was possible to describe a disease which originated from defective mitochondrial DNA and a debate had started as to the use and value of gene therapy. More information about this can be found in Chapter 10.

Bibliography

Chromatin structure

Adachi, Y., Luke, M. and Laemmli, U.K. (1991) Chromosome assembly in vitro: topoisomerase II is required for condensation. *Cell* **64**, 137–148.

Adams, R.L.P., Burdon, R.H., Campbell, A.M., *et al.* (1981) *The Biochemistry of the Nucleic Acids*, 9th edn, Chapman & Hall, London.

Burkholder, G. (1988) The analysis of chromosome organization by experimental manipulation, in *Chromosome Structure and Function* (eds J.P. Gustavson and R. Appels), Plenum Press, New York.

Burns, G.W. and Bottino, P.J. (1989) *The Science of Genetics*, 6th edn, Macmillan Publishing Company, USA.

Earnshaw, W.C. (1988) Mitotic chromosome structure. *BioEssays* **9**, 147–150.

Filipski, J., Leblanc, J., Youdale, T., et al. (1990) Periodicity of DNA folding in higher order chromatin structures. *EMBO J.* **9**, 1319–1327.

Flavell, R.B., Bennett, M.D., Smith, J.B., Smith, D.B. (1974) Genome size and the proportion of repeated nucleotide sequence DNA in plants. *Biochem. Genet.* **12**, 257–271.

Gasser, S.M. and Laemmli, U.K. (1987) A glimpse at chromosomal order. *Trends Genet.* **3**, 16–22.

Gerdes, M.G., Carter, K.C., Moen Jr, P.T. and Bentley-Lawrence, J. (1994) Dynamic changes in the higher-level chromatin organisation of specific sequences revealed by *in situ* hybridization to nuclear halos. *J. Cell. Biol.* **126**, 289–304.

Jackson, D.A. (1991) Structure–function relationships in eukaryotic nuclei. *BioEssays* **13**, 1–12.

Laemmli, U.K., Cheng, S.M., Adolph, K.W. *et al.* (1977) Metaphase chromosome structure: the role of nonhistone proteins. *Cold Spring Harbor Symp. Quant. Biol.* **XLII**, 351–360.

Manuelidis, L. (1990) A view of interphase chromosomes. *Science* **250**, 1533–1540.

Manuelidis, L. and Borden, J. (1988) Reproducible compartmentaliza-tion of individual chromosome domains in human CNS cells revealed by *in situ* hybridization and three-dimensional reconstruc-tion. *Chromosoma* **96**, 397–410.

Manuelidis, L. and Chen, T.L. (1990) A unified model of eukaryotic chromosomes. *Cytometry* **11**, 8–25.

Peterson, C.L. (1994) The SMC family: novel motor proteins for chromosome condensation. *Cell.* **79**, 389–392.

Reeves, R. (1992) Chromatin changes during the cell cycle. *Curr. Opin. Cell. Biol.* **4**, 413–423.

Watson, J.D. and Crick, F.H.C. (1953) Molecular structure of nucleic acids: a structure for deoxyribose nucleic acid. *Nature* **171**, 737–738.

Chromosome replication

Adams, R.L.P., Burdon, R.H., Campbell, R.H., *et al.* (1981) *The Bio-chemistry of the Nucleic Acids*, Chapman & Hall, London.

Adams, R.L.P., Knowler, J.T. and Leader, D.P. (1992) *The Biochemistry of the Nucleic Acids*, 11th edn, Chapman & Hall, London.

Benbow, R.M., Zhao, J. and Larson, D.D. (1992) On the nature of the origins of DNA replication in eukaryotes. *BioEssays* **14**, 661–670.

DePamphilis, M.L. (1993) Origins of DNA that function in eukaryotic cells. *Curr. Opin. Cell. Biol.* **5**, 434–441.

Gerdes, M.G., Carter, K.C., Moen Jr, P.T. and Bentley-Lawrence, J. (1994) Dynamic changes in the higher-level chromatin organisation of specific sequences revealed by *in situ* hybridization to nuclear halos. *J. Cell. Biol.* **126**, 289–304.

Gruss, C. and Sogo, J.M. (1992) Chromatin replication. *BioEssays* **14**, 1–8.

Hamlin, J.L. (1992) Mammalian origins of replication. *BioEssays* **14**, 651–659.

Taylor, J.H., Woods, P.S. and Hughes, W.L. (1957) The organization and duplication of chromosomes as revealed by autoradiographic studies using tritium-labeled thymidine. *Proc. Natl. Acad. Sci. USA* **43**, 122–128.

Watson, J.D. and Crick, F.H.C. (1953) Genetic implications of the structure of deoxyribonucleic acid. *Nature* **171**, 964–967.

Heterochromatin

Blumberg, B.D., Shulkin, J.D., Rotter, J.I., Mohandas, T. and Kaback, M. (1982) Minor chromosomal variants and major chromosomal anomalies in couples with recurrent abortion. *Am. J. Hum. Genet.* **34**, 948–960.

Bobrow, M. (1985) Heterochromatic chromosome variation and reproductive failure. *Exp. Clin. Immunogenet.* **2**, 97–105.

Flavell, R.B. (1986) Repetitive DNA and chromosome evolution in plants. *Philos. Trans. R. Soc.* **312**, 227–242.

John, B. (1988) The biology of heterochromatin, in *Heterochromatin* (ed. R.S. Verma), Cambridge University Press, Cambridge, pp.1–128.

John, B. and Miklos, G.L.G. (1987) *The Eukaryote Genome in Development and Evolution*, Allen & Unwin, London.

Kit, S. (1961) Equilibrium centrifugation in density gradients of DNA preparations from animal tissues. *J. Mol. Biol.* **3**, 711–716.

Maes, A., Staessen, C., Hens, L., *et al.* (1983) C heterochromatin variation in couples with recurrent early abortions. *J. Med. Genet.* **20**, 350–356.

Mahan, J.T. and Beck, M.L. (1986) Heterochromatin in mitotic chromosomes of the *virilis* group of *Drosophila. Genetics* **68**, 113–118.

Nielsen, J. and Friedrich, U. (1972) Length of the Y chromosome in criminal males. *Clinical Genetics* **3**, 281–285.

Petkovic, I., Nakic, M., Tiefenbach, A., *et al.* (1987) Heterochromatic segment length of Y chromosome in 55 boys with malignant disease. *Cancer Genet. Cytogenet.* **25**, 351–353.

Rodriguez-Gomez, M.T., Martin-Sempere, M.J. and Abrisqueta, J.A. (1987) C-band length variability and reproductive wastage. *Hum. Genet.* **75**, 56–61.

Ronchetti, E., Crovella, S., Rumpler, Y.M., *et al.* (1993) Genome size and qualitative and quantitative characteristics of C-heterochromatin DNA in *Eulemur* species and in a viable hybrid. *Cytogenet. Cell Genet.* **63**, 1–5.

Schweizer, D. and Loidl, J. (1987) A model for heterochromatin dispersion and the evolution of C-band patterns. *Chrom. Today* **10**, 61–74.

Telomeres

Allshire, R.C., Gosden, J.R., Cross, S.H. *et al.* (1988) Telomeric repeat from *T. thermophila* cross-hybridizes with human telomeres. *Nature* **332**, 656–659.

Boeke, J.D. (1990) Reverse transcriptase, the end of the chromosome, and the end of life. *Cell* **61**, 193–195.

Broccoli, D. and Cooke, H. (1993) Aging, healing and the metabolism of telomeres. *Am. J. Hum. Genet.* **52**, 657–660.

Brown, W.R.A. (1992) Telomerase and chromosome healing. *Curr. Biol.* **2**, 127–129.

Greider, C.W. (1991) Chromosome first aid. *Cell* **67**, 645–647.

Kipling D. and Cooke, H.J. (1992) Beginning or end? Telomere structure, genetics and biology. *Hum. Mol. Genet.* **1**, 3–6.

Meyne, J., Baker, R.J., Hobart, H.H., *et al.* (1990) Distribution of non-telomeric sites of the (TTAGGG)$_n$ telomeric sequence in vertebrate chromosomes. *Chromosoma* **99**, 3–10.

Morin, G.B. (1991) Recognition of a chromosome truncation site associated with alpha thalassaemia by human telomerase. *Nature* **353**, 454–456.

Moyzis, R.K., Buckingham, J.M., Scott-Cram, L., *et al.* (1988) A highly conserved repetitive DNA sequence, (TTAGGG)$_n$, present at the telomeres of human chromosomes. *Proc. Natl. Acad. Sci. USA* **85**, 6622–6626.

Muller, F., Wicky, C., Spicher, A. and Tobler, H. (1991) New telomere formation after developmentally regulated chromosomal breakage during the process of chromatin diminution in *Ascaris lumbricoides. Cell* **67**, 815–822.

Nurnberg, P., Thiel, G., Weber, F. and Epplen, J.T. (1993) Changes of telomere lengths in human intracranial tumours. *Hum. Genet.* **91**, 190–192.

Saltman, D., Morgan, R., Cleary, M.L. and De Lange, T. (1993) Telomeric structure in cells with chromosome end associations. *Chromosoma* **102**, 121–128.

Schwarzacher, T. and Heslop-Harrison, J.S. (1991) *In situ* hybridisation to plant telomeres using synthetic oligomeres. *Genome* **34**, 317–323.

Stoll, S., Zirlik, T., Maerker, C. and Lipps, H.J. (1993) The organisation of internal telomeric repeats in the polytene chromosomes of the hypotrichous ciliate *Stylonychia lemnae. Nucleic Acids Res.* **12**, 1783–1788.

Tobler, H., Etter, A. and Muller, F. (1992) Chromatin diminution in nematode development. *Trends Genet.* **8**, 427–432.

Van Der Ploeg, L.H.T., Gottesdiener, K. and Lee, M. G-S. (1992) Antigenic variation in African trypanosomes. *Trends Genet.* **8**, 452–457.

Vaziri, H., Schachter, F., Uchida, I., *et al.* (1993) Loss of telomeric DNA during aging of normal and trisomic 21 human lymphocytes. *Am. J. Hum. Genet.* **52**, 661–667.

Wright, W.E. and Shay J.W. (1992) Telomere positional effects and the regulation of cellular senescence. *Trends Genet.* **8**, 193–197.

Centromeres and kinetochores

Alfenito, M.R. and Birchler, J.A. (1993) Molecular characterization of a maize B chromosome centric sequence. *Genetics* **135**, 589–597.

Cooper, K.F. and Tyler-Smith, C. (1992) The putative centromere forming sequence lambda CM8 is a single copy sequence and is not a component of most human centromeres. *Hum. Mol. Genet.* **1**, 753–754.

Earnshaw, W.C., Ratrie, H. and Stetten, G. (1989) Visualisation of centromere proteins CENP-B and CENP-C on a stable dicentric chromosome in cytological spreads. *Chromosoma* **98**, 1–12.

Earnshaw, W.C., Saitoh, H., Tomkiel, J.E., *et al.* (1993) Molecular cloning and characterization of human centromeric autoantigen CENP-C: a component of the inner kinetochore plate. *Chrom. Today* **11**, 23–34.

Erickson, J.M., Rushford, C.L., Dorney, D.J., *et al.* (1981) Structure and variation of human ribosomal DNA. *Gene* **16**, 1–9.

Haas, O.A. (1990) Centromeric heterochromatin instability of chromosome 1, 19 and 16 in variable immunodeficiency syndrome – a virus induced phenomenon? *Hum. Genet.* **85**, 244.

Johnson, D.H., Kroisel, P.M., Klapper, H.J. and Rosenkranz, W. (1992) Microdissection of a human marker chromosome reveals its origin and a new family of centromeric repetitive DNA. *Hum. Mol. Genet.* **1**, 741–747.

Lacadena, J.R., Cermeno, M.C., Orellana, J. and Santos, J.L. (1988) Nucleolar competition in Triticeae, in *Kew Chromosome Conference III* (ed. P.E. Brandham), HMSO, London, pp. 151–165.

McGill, N.I., Fantes, J. and Cooke, H. (1992) Lambda CM8, a human sequence with putative centromeric function does not map to the centromere but is present in one to two copies at 9qter. *Hum. Molec. Genet.* **1**, 749–751.

Miller, O.J. (1981) Nucleolar organisers in mammalian cells. *Chrom. Today* **7**, 64–73.

Murakami, S., Hatsumoto, T., Niwa, O. and Yanagida, M. (1991) Structure of the fission yeast centromere Cen3: direct analysis of the reiterated inverted region. *Chromosoma* **101**, 214–221.

Murray, A.W. and Mitchison, T.J. (1994) Kinetochores pass the IQ test. *Curr. Biol.* **4**, 38–41.

Pluta, A.F., Cooke, C.A. and Earnshaw, W.C. (1990) Structure of the human centromere at metaphase. *Trends Biochem. Sci.* **15**, 181–185.

Rattner, J.B. and Linn, C.C. (1985) Centromere organisation in chromosomes of the mouse. *Chromosoma* **92**, 325–329.

Schmickel, R.D. (1987) The molecular organization of the human ribosomal gene. *Chrom. Today* **9**, 242–251.

Skibbens, R., Skeen, V.P. and Salmon, E.D. (1993) Directional stability of kinetochore mobility during chromosome congression and segregation in mitotic newt lung cells: a push–pull mechanism. *J. Cell Biol.* **122**, 859–975.

Tartof, K.D. (1973) Regulation of ribosomal RNA gene multiplicity in *Drosophila melanogaster. Genetics* **73**, 57–71.

Tyler-Smith, C., Oakey, R., Larin, Z. *et al.* (1993) Localization of DNA sequences required for human centromere function through an analysis of rearranged Y chromosomes. *Nature Genet.* **5**, 368–375.

Willard, H. (1990) Molecular cytogenetics of centromeres of human chromosomes. *Chrom. Today* **10**, 47–60.

Willard, H. (1992) Centromeres-primary constrictions are primarily complicated. *Hum. Mol. Genet.* **1**, 667–668.

Zhang, A., Lin., M.S. and Wilson, M.G. (1987) Effect of C-band heterochromatin on centromere separation. *Hum. Hered.* **37**, 285–289.

Lampbrush chromosomes

Flemming, W. (1882) *Zellsubstanz, Kern und Zelltheilung*, Vogel, Leipzig.

Gall, J.G. (1954) Lampbrush chromosomes from oocyte nuclei of the newt. *J. Morphol.* **94**, 283–352.

Green, D.M. (1988) Cytogenetics of the endemic New Zealand frog *Leiopelma hochstetteri*: extraordinary supernumerary chromosome variation and a unique sex chromosome system. *Chromosoma* **97**, 55–70.

Green, D.M., Kezer, J. and Nussbaum, R.A. (1987) Supernumerary chromosome variation and heterochromatin distribution in the endemic New Zealand frog *Leiopelma hochstetteri. Chromosoma* **95**, 339–344.

Macgregor, H.C. and Andrews, C. (1977) The arrangement and transcription of middle repetitive DNA sequences on lampbrush chromosomes of *Triturus. Chromosoma* **63**, 109–126.

Nussbaum, R.A. (1991) Cytotaxonomy of Caecilians, in *Amphibian*

Cytogenetics and Evolution (eds D.M. Green and S.K. Sessions), Academic Press, London.

Shang, X.M. and Wang, W.L. (1991) DNA amplification, chromatin variations, and polytene chromosomes in differentiating cells of common bread wheat *in vitro* and roots of regenerated plants *in vivo*. *Genome* **34**, 799–809.

Sims, S.H., Macgregor, H.C., Pellat, P.A. and Horner, H.A. (1984) Chromosome 1 in crested and marbled newts (*Triturus*). An extraordinary case of heteromorphism and independent chromosome evolution. *Chromosoma* **89**, 169–185.

Zacharopoulou, A., Bourtzis, K. and Kerremans, P. (1991) A comparison of polytene chromosomes in salivary glands and orbital trichogen cells in *Ceratitis capitata*. *Genome* **34**, 215–219.

Double minutes and homogeneously staining regions

Benner, S.E., Wahl, G.M. and Von-Hoff, D.D. (1991) Double minute chromosomes and homogeneously staining regions in tumors taken directly from patients versus human tumor cell lines. *Anti-Cancer Drugs* **2**, 11–25.

Bertino, J.R., Srimatkandada, S., Engel, D., *et al.* (1982) Gene amplification in a methotrexate-resistant human leukemia line, K-562, in *Gene Amplification*, (ed. R.T. Schimke), (Cold Spring Harbor Laboratory Press, Cold Spring Harbor, NY, pp. 23–27.

Schimke, R.T. (ed.) (1982) *Gene Amplification*, Cold Spring Harbor Laboratory Press, Cold Spring Harbor, NY.

Supernumerary segments and B chromosomes

Alvarez, M.T., Forminaya, A. and Perez de la Vega, M. (1991) A possible effect of B-chromosomes on metaphase I homologous chromosome association in rye. *Heredity* **67**, 123–128.

Carlson, W.R. and Roseman, R.R. (1992) A new property of the maize B chromosome. *Genetics* **131**, 211–223.

Green, D.M. (1988) Cytogenetics of the endemic New Zealand frog, *Leiopelma hochstetteri*: extraordinary supernumerary chromosome variation and a unique sex-chromosome system. *Chromosoma* **97**, 55–70.

Green, D.M. (1990) Muller's ratchet and the evolution of supernumerary chromosomes. *Genome* **33**, 818–824.

Green, D.M. (1991) Supernumerary chromosomes in amphibians, in *Amphibian Cytogenetics and Evolution* (eds D.M. Green and S.K. Sessions), Academic Press, London, pp. 333–357.

Holmes, D.S. and Bougourd, S.M. (1991) B-chromosome selection in *Allium schoenoprasum*. I. Experimental selection. *Heredity* **67**, 117–122.

Jones, G.H., Albini, S.M. and Whitehorn, J.A.F. (1991) Ultrastructure of meiotic pairing in B-chromosomes of *Crepis capillaris*. II. 4 B pollen mother cells. *Chromosoma* **100**, 193–202.

Jones, R.N. and Rees, H. (1982) *B-chromosomes*. Academic Press, New York.

Parker, J.S., Jones, G.H., Edgar, L.A., and Whitehouse, C. (1990) The population cytogenetics of *Crepis capillaris*. III. B-chromosome effects on meiosis. *Heredity* **64**, 377–385.

Rees, H., and Jones, R.N. (1977) *Chromosome Genetics*, Edward Arnold, London.

Romera, F., Jimenez, M.M. and Peurtas, M.J. (1991) Genetic control of the rate of transmission of rye B chromosomes. I. Effects in 2B x 0B crosses. *Heredity* **66**, 61–65.

Sandery, M.J., Forster, J.W., Blunden, R., and Jones, R.N. (1990) Identification of a family of repeated sequences on the rye B-chromosome. *Genome* **33**, 908–913.

Staub, R.W. (1987) Leaf striping correlated with the presence of B chromosomes in maize. *J. Hered.* **78**, 71–74.

Wilby, A.S. and Parker, J.S. (1988) The supernumerary segment systems of *Rumex acetosa*. *Heredity*, **60**, 109–117.

Artificial chromosomes

Anand, R. (1992) Yeast artificial chromosomes (YACs) and the analysis of complex genomes. *Trends Biotechnol.* **10**, 35–40.

Capecchi, M.R. (1993) Yacs to the rescue. *Nature* **362**, 205–206.

Clarke, L. and Carbon, J. (1980) Isolation of a yeast centromere and construction of functional small circular chromosomes. *Nature* **287**, 504–509.

Featherstone, T. and Huxley, C. (1993) Extrachromosomal maintenance and amplification of yeast artificial chromosome DNA in mouse cells. *Genomics* **17**, 267–278.

Guerrini, A.M., Ascenzioni, F., Pisani, G., *et al.* (1990) Cloning a fragment from the telomere of the long arm of human chromosome 9 in a YAC vector. *Chromosoma* **99**, 138–142.

Huxley, C. (1994) Mammalian artificial chromosomes: a new tool for gene therapy. *Gene Therapy* **1**, 7–12.

Zabel, P., Meyer, D., Stolpe, O., *et al.* (1985) Towards the construction of artificial chromosomes for tomato, in *Molecular Form and Function of the Plant Genome* (eds G.S.V. Vloten-Doting, G.S.V. Groot and T.C. Hall), Plenum Press, New York, pp. 609–624.

Banding techniques – reviews

Allen, T., Jack, E., Harrison, C. and Claugher, D. (1986) Scanning electron microscopy of human metaphase chromosomes. *Scan. Electron Microsc.* **1**, 301–308.

Allen, T., Jack, E., Harrison, C., et al. (1985) *Science of Biological Specimen Preparation*, SEM, Chicago, pp. 299–307.

Barnes, I.C.S. and Maltby, E.L. (1986) Prometaphase chromosome analysis as a routine diagnostic technique. *Clin. Genet.* **29**, 378–383.

Bickmore, W.A. and Sumner, A.T. (1989) Mammalian chromosome banding – an expression of genome organisation. *Trends Genet.* **5**, 144–148.

Comings, D.E. (1978) Mechanisms of chromosome banding and implications for chromosome structure. *Annu. Rev. Genet.* **12**, 25–46.

Evans, H.J. (1988) Mutation as a cause of genetic disease. *Phil. Trans. R. Soc. Lond. B* **319**, 115–130.

Holmquist, G. (1992) Chromosome bands, their chromatin flavors and their functional features. *Am. J. Hum. Genet.* **51**, 17–37.

Rooney, D. and Czepulkowski, B. (1986) *Human Cytogenetics, a Practical Approach*, IRL Press, Oxford.

Sumner, A. (1990) *Chromosome Banding*, Allen & Unwin, London.

G-banding

Buys, C.H.C.M., Aanstoot, G.H. and Nienhaus, A.J. (1984) The giemsa-11 technique for species-specific chromosome differentiation. *Histochemistry* **81**, 465–468.

Drewry, A. (1982) G-banded chromosomes in *Pinus resinosa. J. Hered.* **73**, 305–306.

Drouin, R., Lemieux, N. and Richer, C.-L. (1991) Chromosome condensation from prophase to late metaphase: relationship to chromosome bands and their replication time. *Cytogenet. Cell Genet.* **57**, 91–99.

Greilhuber, J. (1977) Why plant chromosomes do not show G-bands. *Theor. Appl. Genet.* **50**, 121–124.

Holmquist, G., Gray, M., Porter, T. and Jordan, J. (1982) Characterization of Giemsa dark- and light-band DNA. *Cell* **31**, 121–129.

Kakeda, K., Yamagata, H., Fukui, K., *et al.* (1990) High resolution bands in maize chromosomes by G-banding methods. *Theor. Appl. Genet.* **80**, 265–272.

Savage, J.R.K. (1977) Assignment of aberration breakpoints in banded chromosomes. *Nature* **270**, 513–514.

Sumner, A.T. (1980) Dye binding mechanisms in G-banding of chromosomes. *J. Microsc.* **119**, 397–400.

C-banding

Braekeleer, M., Keushnig, M. and Lin, C.C. (1986) A high-resolution C-banding technique. *Can. J. Genet. Cytol.* **28**, 317–322.

Jack, E., Harrison, C., Allen, T. and Harris, R. (1985) The structural basis for C-banding: a scanning electron microscopy study. *Chromosoma* **91**, 363–368.

Linde-laursen, I. (1975) Giemsa C-banding of the chromosomes of Emir barley. *Hereditas* **81**, 285–289.

Pijnacker, L.P. and Ferwerda M.A. (1984) Giemsa C-banding of potato chromosomes. *Can. J. Genet. Cytol.* **26**, 415–419.

Verma, R.S. (1988) *Heterochromatin*, Cambridge University Press, Cambridge.

Wall, W.J. and Butler, L.J. (1989) Classification of Y chromosome polymorphisms by DNA content and C-banding. *Chromosoma* **97**, 296–300.

Fluorescent banding

Anderson, O. and Ronne, M. (1986) Effects of ethidium bromide and bisbenzimide (Hoechst 33258) on human lymphocyte chromosome structure. *Hereditas* **105**, 269–272.

Bella, J.L., Garcia de la Vega, C., Lopez-fernandez, C. and Gosalvez, J. (1986) Changes in acridine orange binding and its use in the characterisation of heterochromatic regions. *Heredity* **57**, 79–83.

Holmquist, G. (1979) The mechanism of C-banding: depurination and B elimination. *Chromosoma* **72**, 225–240.

Lichtenberger, M.J. (1983) Quick and reversible staining methods for G-and Q-bands of chromosomes. *Stain Technol.* **58**, 185–188.

Sumner, A.T. (1986) Mechanisms of quinacrine binding and fluorescence in nuclei and chromosomes. *Histochemistry* **84**, 566–574.

Sumner, A. and Evans H.J. (1973) Mechanisms involved in the banding of chromosomes with quinacrine and Giemsa. *Exp. Cell Res.* **81**, 223–236.

R-banding

Bird, A.P. (1987) CpG islands as gene markers in the vertebrate nucleus. *Trends Genet.* **3**, 342–347.

Craig, J.M. and Bickmore, W.A. (1994) The distribution of CpG islands in mammalian chromosomes. *Nature Genet.* **7**, 376–382.

Hirsch B., Mack, R. and Arthur, D. (1987) Sequential G- to R-banding for high resolution chromosome analysis. *Hum. Genet.* **76**, 37–39.

Jack, E., Harrison, C., Allen, T. and Harris, R. (1986) A structural basis for R- and T-banding: a scanning electron microscopy study. *Chromosoma* **94**, 395–402.

Scheres, J.M., Hustinx, T.W.J. and Merkx, G.F.M. (1980) Normarski-optical studies of human chromosomes R-banded with barium hydroxide. *Hum. Genet.* **53**, 255–259.

BrdU staining

Aghamohammadi, S.Z. and Savage, J.R.K. (1990) BrdU pulse/reverse staining protocols for investigating chromosome replication. *Chromosoma* **99**, 76–82.

Hellmer, A., Voiculescu, I. and Schempp, W. (1991) Replication banding studies in two cyprinid fishes. *Chromosoma* **100**, 524–531.

Krawczun, M.S., Camargo, M. and Cervenka, J. (1986) Patterns of BrdU incorporation in homogeneously staining regions and double minutes. *Cancer Genet. Cytogenet.* **21**, 257–265.

Pijnacker, L.P., Walch K. and Ferwerda M.A. (1986) Behaviour of chromosomes in potato leaf tissue cultured *in vitro* as studied by BrdC–Giemsa labelling. *Theor. Appl. Genet.* **72**, 833–839.

RE-banding

Bianchi, N.O., and Bianchi, M.S. (1987) Analysis of the eukaryotic chromosome organisation with restriction endonucleases, in *Cytogenetics* (eds G. Obe and A. Basler), Springer, Heidelberg, pp. 280–299.

Burkholder, G.D. (1989) Morphological and biochemical effects of endonucleases on isolated mammalian chromosomes *in vitro*. *Chromosoma* **97**, 347–355.

Lozano, R., Jamilena, M., Ruiz Rejon, C. and Ruiz Rejon, M. (1990) Characterization of the chromatin of some liliaceous species after digestion with restriction endonucleases and sequential Giemsa, fluorochrome and silver staining. *Heredity* **64**, 185–195.

Lozano, R., Ruiz Rejon, M. and Ruiz Rejon, R. (1991) An analysis of COHO salmon chromatin by means of C-banding, AG- and fluorochrome staining, and *in situ* digestion with restriction endonucleases. *Heredity* **66**, 403–409.

Lozano, R., Sentis, C., Fernandez-Piqueras, J. and Ruiz Rejon, M. (1991) *In situ* digestion of satellite DNA of *Scilla siberica*. *Chromosoma* **100**, 439–442.

Sanchez, L., Martinez, P., Bouza, C. and Vinas, A. (1991) Chromosomal heterochromatin differentiation in *Salmo trutta* with restriction enzymes. *Heredity* **66**, 241–249.

Schmid, M. and de Almeida, C.G. (1988) Chromosome banding in amphibia. XII. Restriction endonuclease banding. *Chromosoma* **96**, 283–290.

Smith, H.O. and Nathan, D. (1973) A suggested nomenclature for bacterial host modification and restriction systems and their enzymes. *J. Mol. Biol.* **81**, 419–423.

In situ *hybridization banding*

Greilhuber, J. (1977) Why plant chromosomes do not show G-bands. *Theor. Appl. Genet.* **50**, 121–124.

Itoh, K., Iwabuchi, M. and Shimamoto, K. (1991) *In situ* hybridization with species-specific DNA probes gives evidence for asymmetric nature of *Brassica* hybrids obtained by X-ray fusion. *Theor. Appl. Genet.* **81**, 356–362.

Korenberg, J.R. and Rykowski, M.C. (1988) Human genome organisation: Alu, Lines, and the molecular structure of metaphase chromosome bands. *Cell* **53**, 391–400.

Lane, M.J., Waterbury, P.G., Carrol, W.T. *et al.* (1992) Variation in genomic Alu repeat density as a basic for rapid construction of low resolution physical maps of human chromosomes. *Chromosoma* **101**, 349–357.

Lapitan, N.L.V., Ganal, M.W. and Tanksley, S.D. (1989) Somatic chromosome karyotype of tomato based on *in situ* hybridization of the TGRI satellite repeat. *Genome* **32**, 992–998.

D-banding

Burkholder, G.D. and Weaver, M.G. (1975) Differential accessibility of DNA in extended and condensed chromatin to pancreatic DNase I. *Exp. Cell Res.* **92**, 518–522.

Elgin, S.C.R. (1981) DNAase I-hypersensitive sites of chromatin. *Cell* **27**, 413–415.

Kerem, B.-S., Goitein, R., Diamond, G. *et al.* (1984) Mapping of DNAase I sensitive regions on mitotic chromosomes. *Cell* **38**, 493–499.

Cell division

Albini, S.M. and Jones, G.H. (1984) Synaptonemal complex-associated centromeres and recombination nodules in plant meiocytes prepared by an improved surface-spreading technique. *Experimental Cell Research* **155**, 588–592.

Ault, J.G. and Rieder, C.L. (1994) Centrosome and kinetochore movement during mitosis. *Curr. Opin. Cell. Biol.* **6**, 41–49.

Bennett, M.D. (1971) The duration of meiosis. *Proc. R. Soc. Lond. B.* **178**, 277–299.

Carpenter, A.T.C. (1987) Gene conversion, recombination nodules, and the initiation of meiotic synapsis. *BioEssays* **6**, 232–236.

Connor, J.M. and Ferguson-Smith, M.A. (1991) *Essential Medical Genetics*, Blackwell Scientific Publications, Oxford.

Earnshaw, W.C. and Tomkiel, J.E. (1992) Centromere and kinetochore structure. *Curr. Opin. Cell. Biol.* **4**, 86–93.

Endow, S.A. (1993) Chromosome distribution, molecular motors and the claret protein. *Trends Genet.* **9**, 52–55.

Evans, G.M. (1988) Genetic control of chromosome pairing in polyploids, in *Kew Chromosome Conference III* (ed. P.E. Brandham) HMSO, London, pp. 253–260.

Gillies, C.B. (1984) The synaptonemal complex in higher plants. *CRC Crit. Rev. Plant Sci.* **2**, 81–116.

Gorbsky, G.J. (1992) Chromosome motion in mitosis. *BioEssays* **14**, 73–80.

Handel, M.A. and Hunt, P.A. (1992) Sex-chromosome pairing and activity during mammalian meiosis. *BioEssays* **14**, 817–822.

Hastings, P.J. (1987) Models of heteroduplex formation, in *Meiosis* (ed. P.B. Moens), Academic Press, New York, pp. 139–156.

Holliday, R. (1964) A mechanism for gene conversion in fungi. *Genet. Res.* **5**, 282–304.

Jones, G.H. and Albini, S.M. (1988) Meiotic roles of nodule structures in zygotene and pachytene nuclei of angiosperms, in *Kew Chromosome Conference III* (ed. P.E. Brandham), HMSO, London, pp. 323–330.

Murray, A.W. and Mitchison, T.J. (1994) Kinetochores pass the IQ test. *Curr. Biol.* **4**, 38–41.

Roeder, G.S. (1990) Chromosome synapsis and genetic recombination. *Trends Genet.* **6**, 385–389.

Singh, R.J. (1993) *Plant Cytogenetics,* CRC Press, Boca, Raton, FL.

Chromosome organization within the nucleus

Bennett, M.D. (1983) The spatial distribution of chromosomes, in *Kew Chromosome Conference II*, (eds P.E. Brandham and M.D. Bennett), HMSO, London, pp. 71–79.

Bennett, M.D. (1988) Parental genome separation in F1 hybrids between grass species, in *Kew Chromosomes Conference III* (eds P.E. Brandham and M.D. Bennett), HMSO, London, pp. 195–208.

Borden, J. and Manuelidis, L. (1988) Movement of the X chromosome in epilepsy. *Science* **242**, 1687–1691.

Carter, K.C., Taneja, K.L. and Lawrence, J.B. (1991) Discrete nuclear domains of poly (A) RNA and their relationship to the functional organisation of the nucleus. *J. Cell. Biol.* **115**, 1191–1202.

Endow, S.A. (1993) Chromosome distribution, molecular motors and the claret protein. *Trends Genet.* **9**, 52–55.

Heslop-Harrison, J.S. (1983) Chromosome disposition in *Aegilops umbellulata*, in *Kew Chromosome Conference II* (eds P.E. Brandham and M.D. Bennett), HMSO, London, pp. 63–70.

Heslop-Harrison, J.S., Huelskamp, M., Wendroth, S. *et al.* (1988) Chromatin and centromeric structures in interphase nuclei, in *Kew Chromosome Conference III* (ed. P.E. Brandham), HMSO, London, pp. 209–217.

Heslop-Harrison, J.S. and Bennett, M.D. (1990) Nuclear architecture in plants. *Trends Genet.* **6**, 401–405.

Manuelidis, L. (1984) Active nucleolus organisers are precisely positioned in adult nervous system cells but not in neuroectodermal tumour cells. *J. Neuropathol. Exp. Neurol.* **43**, 225–241.

Manuelidis, L. (1985) Individual interphase chromosome domains revealed by *in situ* hybridization. *Hum. Genet.* **71**, 288–293.

Manuelidis, L. (1990) A view of interphase chromosomes. *Science* **250**, 1533–1540.

Sumner, A.T. and Chandley, A.C. (1993) Chromosomes Today, vol 11, Chapman & Hall, London.

Trask, B.J. (1991) Fluorescence *in situ* hybridization: applications in cytogenetics and gene mapping. *Trends Genet.* **7**, 149–154.

Imprinting

Caron, H., Van Sluis, P., Van Hoeve, M. *et al.* (1993) Allelic loss of chromosome 1p36 in neuroblastoma is of preferential maternal origin and correlates with N-*myc* activity. *Nature Genet.* **4**, 187–190.

Cattanach, B.M., and Beechey, C.V. (1990) Chromosome imprinting phenomena in mice and indications in man, in *Chromosomes Today*, Vol.10 (eds K. Fredga, B.A. Kihlman and M.D. Bennett), Unwin Hyman, London, pp. 135–148.

Cheng, J.M., Hiemstra, J.L., Schneider, S.S. *et al.* (1993) Preferential amplification of the paternal allele of the N-*myc* gene in human neuroblastomas. *Nature Genet* **4**, 191–195.

Feinberg, A.P. (1993) Genomic imprinting and gene activation in cancer. *Nature Genet.* **4**, 110–113.

Hall, J.G. (1991) Genomic imprinting. *Curr. Opin. Genet. Dev.* **1**, 34–39.

Heutink, P., Van Der May, A.G.L., Sandkuijl, L.A. *et al.* (1992) A gene subject to genomic imprinting and responsible for hereditary paragangliomas maps to chromosome 11q23–qter. *Hum. Molec. Genet.* **1**, 7–10.

Hulten, M.A. and Hall, J.G. (1990) Proposed meiotic mechanism of genomic imprinting. *Chromosomes Today*, Vol.10 (eds K. Fredga, B.A. Kihlman and M.D. Bennett), Unwin Hyman, London, pp. 157–162.

Kennerknecht, I. (1992) A genetic model for the Prader-Willi syndrome and its implications for Angelman syndrome. *Hum. Genet.* **90**, 91–98.

Laird, C.D., Sved, J., Thorne, J. and Lamb, M. (1990) The X-inactivation imprinting model of the fragile X syndrome: annotated references, 1989, in *Chromosomes Today*, Vol.10 (eds K. Fredga, B.A. Kihlman and M.D. Bennett), Unwin Hyman, London, pp. 163–165.

Leff, S.E., Brannan, C.I., Reed M.L. *et al.* (1992) Maternal imprinting of the mouse Snrpn gene and conserved linkage homology with the human Prader-Willi syndrome region. *Nature Genet.* **2**, 259–264.

Meehan, R., Lewis, J., Jeppesen, P. and Bird, A. (1993) Methylated DNA-binding proteins and chromatin structure, in *Chrom. Today*, Vol. 11, Chapman & Hall, London, 377–389.

Peterson, K. and Sapienza, C. (1993) Imprinting the genome: Imprinted genes, imprinting genes and a hypothesis for their interaction. *Annu. Rev. Genet.* **27**, 7–31.

Reik, W., Collick, A., Norris, M.L., *et al.* (1987) Genomic imprinting determines methylation of parental alleles in transgenic mice. *Nature* **328**, 248–251.

Reik, W., Sasaki, H., Ferguson-Smith, A., *et al.* (1993) Paternal imprinting and epigenetic programming of the mouse genome: long lasting consequences for development and phenotype, in *Chrom. Today*, Vol. 11, Chapman & Hall, London, pp. 367–376.

Sapienza, C., Peterson, A.C., Rossant, J. and Balling, R. (1987) Degree of methylation of transgenes is dependent on gamete of origin. *Nature* **328**, 251–254.

Wu, X., Hadchouel, M., Farza, H. *et al.* (1990) Analysis of the sex dependent imprinting of a chromosome 13 region using a transgene as a molecular probe. *Chrom. Today* **10**, 149–156.

Mutagenesis

Antonarakis, E.A., Avramopoulos, D., Blouin, J.-L. *et al.* (1993) Mitotic errors in somatic cells cause trisomy 21 in about 4.5% of cases and are not associated with advanced maternal age. *Nature Genet.* **3**, 146–150.

Balcells, L., Swinburn J. and Coupland G. (1991) Transposons as tools for the isolation of plant genes. *Trends Biotechol.* **9**, 31–37.

Cattanach, B.M., Burtenshaw, M.D., Rasberry, C. and Evans, E.P. (1993) Large deletions and other gross forms of chromosome imbalance compatible with viability and fertility in the mouse. *Nature Genet* **3**, 56–61.

Cleary, M.L. (1991) Oncogenic conversion of transcription factors by chromosomal translocations. *Cell* **66**, 619–622.

Coen, E.S., Carpenter, R. and Martin, C. (1986) Transposable elements generate novel spatial patterns of gene expression in *Antirrhinum majus. Cell* **47**, 285–296.

DeGrouchy, J. (1986) Chromosome phylogenies of man, great apes and old world monkeys. *Genetics* **73**, 37–52.

Djabali, M., Selleri, L., Parry, P., *et al.* (1992) A trithorax-like gene is interrupted by chromosome 11q23 translocations in acute leukaemias. *Nature Genet.* **2**, 113–118.

Evans, H.J. (1988) Mutation as a cause of genetic disease. *Philos. Trans. Soc. Lond.* **319**, 325–340.

Ferguson-Smith, M.A. and Yates, J.P.W. (1984) Maternal age specific rates for chromosome aberrations and factors influencing them: report of a collaborative European study on 52,965 amniocenteses. *Prenat. Diagn.* **4**, 5–44.

Finnegan, D.J. (1985) Transposable elements in eukaryotes. *Int. Rev. Cytol.* **93**, 281–326.

Flavell, R.B. (1986) Repetitive DNA and chromosome evolution in plants. *Philos. Trans. R. Soc. Lond. B* **312**, 227–242.

Grandbastien, M.-A. (1992) Retroelements in higher plants. *Trends Genet.* **8**, 103–108.

Jones, D.A., Thomas, C.M., Hammond-Kossack, K.E., *et al.* (1994) Isolation of the tomato Cf-9 gene for resistance to *Cladosporium fulvum* by transposon tagging. *Science* **266**, 789–792.

Krawczak, M. and Cooper, D.N. (1991) Gene deletions causing human genetic disease: mechanisms of mutagenesis and the role of the local DNA sequence environment. *Hum. Genet.* **86**, 425–441.

Kumar, A. and Gupta, J.P. (1992) Concentration of chromosomal aberrations on chromosome 3 of *Drosophila nasuta. Heredity* **69**, 263–267.

Lawrence, C.W. (1971) *Cellular Radiobiology*, Institute of Biology Studies in Biology No 30, Edward Arnold, London.

Lorda-Sanchez, I., Binkert, F., Hinkel, K.G., *et al.* (1992) Uniparental origin of sex chromosome polysomies. *Hum. Hered.* **42**, 193–197.

Lucchesi, J.C. and Susuki, D.T. (1968) The interchromosomal control of recombination. *Annu. Rev. Genet.* **2**, 53–86.

Meyn, M.S. (1993) High spontaneous interchromosomal recombination rates in ataxia telangiectasia. *Science* **260**, 1327–1330.

Revel, S.H. (1966) Evidence for a dose-squared term in the dose–response curve for real chromatid discontinuities induced by X-rays and some theoretical consequences thereof. *Mutat. Res.* **3**, 34–53.

Saedler, H. and Nevers, P. (1985) Transposition in plants: a molecular model. *EMBO J.* **4**, 585–590.

Sankaranarayanan, K. (1993) Ionizing radiation, genetic risk estimation and molecular biology: impact and inferences. *Trends Genet.* **9**, 79–84.

Smyth, D.R. (1991) Dispersed repeats in plant genomes. *Chromosoma* **100**, 355–359.

Sparrow, A.H. (1965) Relationship between chromosome volume and radiation sensitivity in plant cells, in *Cellular Radiation Biology*, Williams & Wilkins, Baltimore, MD.

United Nations Scientific Committee on the Effects of Atomic Radiation (1997), in *Sources and Effects of Ionising Radiation, United Nation, New York*, pp. 425–564.

Verschaeve, L., Domracheva, E.V., Kuznetsov, S.A. and Nechai, V.V. (1993) Chromosome aberrations in inhabitants of Byelorussia: consequence of the Chernobyl accident. *Mutat. Res.* **287**, 253–259.

Vogel, F. (1992) Risk calculations for hereditary effects of ionizing radiation in humans. *Hum. Genet.* **89**, 127–146.

Wall, W.J. (1992) Stability of human fibroblast chromosomes in cultured cells. *Chromatin* **1**, 27–30.

Wolff, D.J. and Schwartz, S. (1993) The effect of Robertsonian translocation on recombination of chromosome 21. *Hum. Mol. Genet.* **2**, 693–699.

Chromosome mapping

Baas, F., Bikker, H., Geurts van Kessel A., *et al.* (1985). The human thyroglobin gene: a polymorphic marker localised distal to C–MYC on chromosome 8 band q24. *Hum. Genet.* **69**, 138–143.

Collins, F.S. (1992) Positional cloning: let's not call it reverse anymore. *Nature Genet.* **1**, 3–6.

Dauwerse, J.G., Wiegant, J., Raap, A.K., *et al.* (1992) Multiple colours by fluorescence *in situ* hybridization using radio-labelled DNA probes create a molecular karyotype. *Hum. Mol. Genet.* **1**, 593–598.

Green, D.K. (1990) Analysing and sorting human chromosomes. *J. Microsc.* **159**, 237–244.

James, M.R., Richard III, C.W., Schott J.-J., *et al.* (1994) A radiation hybrid map of 506 STS markers spanning human chromosome 11. *Nature Genet.* **8**, 70–76.

Lander, E.S. and Schork, N.J. (1994) Genetic dissection of complex traits. *Science* **265**, 2037–2048.

Parra, I. and Windle, B. (1993) High resolution visual mapping of stretched DNA by fluorescent hybridization. *Nature Genet.* **5**, 17–21.

Prince, J.P., Pochard, E. and Tanksley, S.D. (1993) Construction of a molecular linkage map of pepper and a comparison of synteny with tomato. *Genome* **36**, 404–417.

Rafalski, J.A., Scott, S.V. and Williams, J.G.K. (1991) RAPD markers – a new technology for genetic mapping and plant breeding. *AgBiotech News Info.* **3**, 645–648.

Riordan, J.R., Rommens, J.M., Kerem, B.-S., *et al.* (1989) Identification of the cystic fibrosis gene: cloning and characterisation of complementary DNA. *Science* **243**, 1066–1073.

Robinson, C (1992) The genome race is on the road. *Trends BioTechnol.* **10**, 1–5.

Rommens, J.M., Iannuzzi, M.C., Kerem B.S., *et al.* (1989) Identification of the cystic fibrosis gene: chromosome walking and jumping. *Science* **243**, 1059–1065.

Schlessinger, D. (1990) Yeast artificial chromosomes: tools for mapping and analysis of complex genomes. *Trends Genet.* **6**, 248–258.

Sherman, S.L. (1991) Workshop Report: Combining genetic and physical maps. *Cytogenet. Cell. Genet.* **58**, 1842–1843.

Skolnick, M. (1991) Workshop Report: Attributes of markers on linkage and physical maps. *Cytogenet. Cell. Genet.* **58**, 1839–1840.

Smith, C.L., Lawrence, S.K., Gillespie G.A., *et al.* (1987) Strategies for mapping and cloning macroregions of mammalian genomes. *Methods Enzymol.* **151**, 461–489.

Tanke, H.J. and van der Keur, M. (1993) Selection of defined cell types by flow-cytometric cell sorting. *Trends. BioTechnol.* **11**, 55–61.

Tanksley, S.D., Young, N.D., Paterson, A.H. and Bonierbale, M.W. (1989) RFLP mapping in plant breeding: new tools for an old science. *Biotechnology* **7**, 257–264.

Tanksley, S.D., Ganal, M.W., Prince, J.P., *et al.* (1992) High density molecular linkage maps of the tomato and potato genomes. *Genetics* **132**, 1141–1160.

Trask, B.J. (1991) Fluorescence *in situ* hybridization: applications in cytogenetics and gene mapping. *Trends Genet.* **7**, 149–154.

Walter, M.A. and Goodfellow, P.N. (1993) Radiation hybrids: irradiation and fusion gene transfer. *Trends Genet.* **9**, 352–356.

Whitkus, R., Doebley, J. and Lee, M. (1992) Comparative genome mapping of sorgum and maize. *Genetics* **132**, 1119–1130.

Williamson, R. (1993) From genome mapping to gene therapy. *Trends BioTechnol.* **11**, 159–161.

Sex chromosomes

Bull, J.J. (1983) *Evolution of Sex Determining Mechanisms*. Benjamin/Cummings, Menlo Park, CA.

Charlesworth, B. (1978) Model for evolution of Y chromosomes and dosage compensation. *Proc. Natl. Acad. Sci. USA.* **75**, 5618–5622.

Jablonka, E. and Lamb, M.J. (1990) The evolution of heteromorphic sex chromosomes. *Biol. Rev.* **65**, 249–276.

Jones, K.W., Singh, L. and Phillips, C. (1983). Conserved nucleotide sequences on sex chromosomes, in *Proceedings of the 5th John Innes Symposium Biological consequences of DNA structure and genome rearrangement* (eds K.F. Chater, C.A. Cullis, D.A. Hopwood, *et al.*) Sinauer, Boston, pp. 265–287.

Muller, H.J. (1964) The relation of recombination to mutational advance. *Mutat. Res.* **1**, 2–9.

Rice, W.R. (1987) Genetic hitchhiking and the evolution of reduced genetic activity of the Y sex chromosome. *Genetics* **116**, 161–167.

Riggs, A.D. (1990) Marsupials and mechanisms of X chromosome inactivation. *Aust. J. Zool.* **37**, 419–441.

Sex chromosomes in mammals

Brown, C.J., Hendrich, B.D., Rupert J.L., *et al.* (1992) The human *Xist* gene: analysis of a 17 kb inactive X-specific RNA that contains conserved repeats and is highly localised in the nucleus. *Cell* **71**, 527–542.

Brown, C.J., Lafreniere, R.G., Powers, V.E., *et al.* (1991) Localisation of the X inactivation centre on the human X chromosome in Xq13. *Nature* **349**, 82–84.

Davies, K.E., Mandel, J.-L., Monaco, A.P., *et al.* (1991) Report of the committee on the genetic constitution of the X chromosome. *Cytogenet. Cell. Genet.* **58**, 853–966.

Disteche, C.M. (1995) Escape from X inactivation in human and mouse. *Trends Genet.* **11**, 17–22.

Fredga, K. (1988) Aberrant chromosomal sex-determining mechanisms in mammals, with special reference to species with XY females. *Philos. Trans. R. Soc. Lond. B* **322**, 83–95.

Goodfellow, P.N. and Lovell-Badge, R. (1993) SRY and sex determination in mammals. *Annu. Rev. Genet.* **27**, 71–92.

Hayman, D.L. (1990) Marsupial cytogenetics. *Aust. J. Zool.* **37**, 331–349.

Lyon, M.F. (1988) The William Allen Memorial Award Address: X-chromosome inactivation and the location and expression of X-linked genes. *Am. J. Hum. Genet.* **42**, 8–16.

Marshall-Graves, J.A. and Watson, J.M. (1991) Mammalian sex chromosomes: evolution of organisation and function. *Chromosoma* **101**, 63–68.

Migeon, B.R. (1994) X chromosome inactivation: molecular mechanisms and genetic consequences. *Trends Genet.* **10**, 230–235.

Norris, D.P., Patel, D., Kay, G.F., *et al.* (1994) Evidence that random and imprinted *Xist* expression is controlled by preemptive methylation. *Cell* **77**, 41–51.

O'Brien, S.J. and Marshall-Graves, J.A. (1991) Report of the committee on comparative gene mapping. *Cytogenet. Cell. Genet.* **58**, 1124–1151.

Sinclair, A.H., Berta, P., Palmer, M.S., *et al.* (1990) A gene from the human sex-determining region encodes a protein with homology to a conserved DNA binding motif. *Nature* **346**, 240–244.

Watson, J.M., Riggs, A. and Marshall-Graves, J.A. (1992) Gene mapping studies confirm the homology between the platypus X and echidna X_1 chromosomes and identify a conserved ancestral monotreme X chromosome. *Chromosoma* **101**, 596–601.

Weissenbach, J. and Goodfellow, P.N. (1991) Report of the committee on the genetic constitution of the Y chromosome. *Cytogenet. Cell. Genet.* **58**, 967–985.

Yen, P.H., Ellison, J., Salido, E.C., *et al.* (1992) Isolation of a new gene from the distal short arm of the human X chromosome that escapes X-inactivation. *Hum. Mol. Genet.* **1**, 47–52.

Sex chromosomes in fish

Ojima, Y. (1983) Fish cytogenetics, in *Chromosomes in the Evolution of Eukaryotic Groups*, (eds A.K. Sharma and A. Sharma), CRC Press, Boca Raton, FL, pp. 111–147.

Price, D.J. (1984) Genetics of sex determination in fishes – a brief review, in *Fish Reproduction* (eds G.W. Potts and R.J. Wooton) Academic Press, London, pp. 77–89.

Sex chromosomes in amphibia

Schmid, M., Nanda, I., Steinlein, C., *et al.* (1991) Sex-determining mechanisms and sex chromosomes in amphibia, in *Amphibian Cytogenetics and Evolution* (eds D.M. Green and S.K. Sessions), Academic Press, New York, pp. 393–430.

Schmid, M. and Haaf, T. (1989) Origin and evolution of sex chromosomes in amphibia: the cytogenetic data, in *Evolutionary Mechanisms in Sex Determination* (ed. S.S. Wachtel), CRC Press, Boca Raton, FL, pp. 38–56.

Sex chromosomes in reptiles

Bull, J.J. (1980) Sex determination in reptiles. *Rev. Biol.* **55**, 3–21.

Bull, J.J. (1989) Evolution and variety of sex-determining mechanisms in amniote vertebrates, in *Evolutionary Mechanisms in Sex Determination*, (ed. S.S. Wachtel), CRC Press, Boca Raton, FL, pp. 57–65.

Deeming, C. and Ferguson, M. (1989) In the heat of the nest. *New Sci.* **1657**, 33–38.

Sex chromosomes in birds

Ansari, H.A., Takagi, N. and Sasaki, M. (1988) Morphological differentiation of sex chromosomes in three species of ratite birds. *Cytogenet. Cell. Genet.* **47**, 185–188.

Baverstock, P.R., Adams, M., Polkinghorne, W. and Gelder, M. (1982) A sex-linked enzyme in birds – Z-chromosome conservation but no compensation. *Nature* **296**, 763–766.

Christidis, L. (1990) *Animal Cytogenetics, Vol. 4, Chordata*, Part 3B: Aves. Gebruder Borntraeger, Berlin.

Nakamura, D., Tiersch, T.R., Douglass, M. and Chandler, R.W. (1990) Rapid identification of sex in birds by flow cytometry. *Cytogenet. Cell. Genet.* **53**, 201–205.

Schmid, M., Enderle, E., Schindler, D. and Schempp, W. (1989) Chromosome banding and DNA replication patterns in bird karyotypes. *Cytogenet. Cell. Genet.* **52**, 139–146.

Sex chromosomes in insects

Animal Cytogenetics Series (various). Gebruder Borntraeger, Berlin.

Franco, M.G., Rubini, P.G. and Vecchi, M. (1982) Sex-determinants and their distribution in various populations of *Musca domestica* L. of Western Europe. *Genet. Res. Camb.* **40**, 279–293.

Traut, W. and Willhoeft, U. (1990) A jumping sex determining factor in the fly *Megaselia scalaris*. *Chromosoma* **99**, 407–412.

Sex chromosomes in plants

Chailakhyan, M. Kh. and Khrianin, V.N. (1987) *Sexuality in Plants and its Hormonal Regulation*. Springer, New York.

Parker, J.S. (1990) Sex chromosomes and sexual differentiation in flowering plants. *Chrom. Today* **10**, 187–198.

Parker, J.S. and Clark, M.S. (1991) Dosage sex-chromosome systems in plants. *Plant Sci.* **80**, 79–92.

Westergaard, M. (1958) The mechanism of sex determination in dioecious flowering plants. *Adv. Genet.* **9**, 217–281.

Evolution and speciation

Aquadro, C.F. (1992) Why is the genome variable? Insights from *Drosophila*. *Trends Genet.* **8**, 355–362.

Birkenmeier, E.H. (1987) Human evolution. *Birth Defects: Original Article Series* **23**, 209–219.

De Grouchy, J. (1987) Chromosome phylogenies of man, great apes and old world monkeys. *Genetica* **73**, 37–52.

Dutrillaux, B. and Couturier, J. (1981) The ancestral karyotype of Platyrrhine M. *Cytogenet. Cell Genet.* **30**, 232–242.

Dutrillaux, B. and Rumpler, Y. (1987) The role of chromosomes in speciation: a new interpretation. *Chrom. Today* **9**, 75–90.

Eldridge, M.D.B. and Close, R.L. (1993) Radiation of chromosome shuffles. *Curr. Opin. Genet. and Dev.* **3**, 915–922.

Erickson, R.P. (1987) Evolution of four human Y chromosomal unique sequences. *J. Mol. Evol.* **25**, 300–307.

Green, D.M. (1990) Muller's ratchet and the evolution of supernumerary chromosomes. *Genome* **33**, 818–824.

Groves, P. (1974) *Horses, Asses and Zebras in the Wild.* David & Charles, Newton Abbot.

Haaf, T. and Schmid, M. (1987) Chromosome heteromorphisms in the gorilla karyotype. *J. Heredity* **78**, 287–292.

Hally, M.K., Rasch, E.M., Mainwairing, H.R. and Bruce, R.C. (1986) Cytophotometric evidence of variation in genome size of desmognathine salamanders. *Histochemistry* **85**, 185–192.

Hayman, D.L., Rofe, R.H. and Sharp, P.J. (1987) Chromosome evolution in marsupials. *Chrom. Today* **9**, 91–102.

Hillis, D.M. (1991) The phylogeny of Amphibians: current knowledge and the role of cytogenetics, in *Amphibian Cytogenetics and Evolution* (eds D. Green and S. Sessions), Academic Press, London, pp. 7–31.

John, B. and Miklos, G. (1988) *The Eukaryote Genome in Development and Evolution.* Allen & Unwin, London.

Kuriyan, P.N., and Narayan, R.K.J. (1988) The distribution and divergence during evolution of families of repetitive DNA sequences in *Lathyrus* species. *J. Mol. Evol.* **27**, 303–310.

Linn, C.C., Sasi, R., Fan, Y.-S. and Chen, Z.-Q. (1991) New evidence for tandem chromosome fusions in karyotype evolution of Asian Muntjacs. *Chromosoma* **101**, 19–24.

Maeda, N., Wu, C.I., Bliska, J. and Reneke, J. (1988) Molecular evolution of intergenic DNA in higher primates: pattern of DNA changes, molecular clock and evolution of repetitive sequences. *Mol. Biol. and Evol.* **5**, 1–20.

Marchant, C.J. (1963) Corrected chromosome numbers for *Spartina* × *townsendii* and its parent species. *Nature* **199**, 929.

Rofe R. and Hayman D. (1985) G-banding evidence for a conserved complement in the Marsupialia. *Cytogenet. Cell Genet.* **39**, 40–50.

Rumpler, Y. and Dutrillaux, B. (1976) Chromosomal evolution in Malagasy Lemurs. I. Chromosomal banding studies in the genuses *Lemur* and *Microcebus. Cytogenet. Cell Genet.* **17**, 268–281.

Sanguthai, O., Sanguthai, S. and Kamemoto, H. (1973) Chromosome doubling of a *Dendrobium* hybrid with colchicine in meristem culture. *Hawaii Orchid J.* **2**, 12–16.

Seuanez, H.N. (1987) The chromosomes of man: evolutionary considerations, in *Cytogenetics* (eds Obe and Basler) Springer, Berlin. pp. 65–89.

Weber, B., Walz, L., Schmid, M. and Scempp, W. (1988) Homoeologic aberrations in human and chimpanzee Y chromosomes: inverted and satellited Y chromosomes. *Cytogenet. Cell Genet.* **47**, 26–28.

Wimber, D.E. and Watrous, S. (1985) *Artificial Induction of Polyploidy in Orchids*, International Centenary Orchid Conference, Royal Horticultural Society, London, pp. 72–79.

Yunis, J.J. and Prakash, O. (1982) The origin of man: A chromosomal pictorial legacy. *Science* **215**, 1525–1530.

Manipulation of whole chromosome sets

Allen, S.K. and Bushek, D. (1992) Large-scale production of triploid oysters *Crassostrea virginica* (Gmelin) using 'stripped' gametes. *Aquaculture* **103**, 241–251.

Beaumont, A.R. and Fairbrother, J.E. (1991) Ploidy manipulation in molluscan shellfish: a review. *J. Shellfish Res.* **10**, 1–18.

Hammatt, N. (1992) Progress in the biotechnology of trees. *World J. Microbiol. Biotechnol.* **8**, 369–377.

Lindsey, K. and Jones, M.G.K. (1989) *Plant Biotechnology in Agriculture.* Open University Press, Milton Keynes.

Maclean, N. and Penman, D. (1990) The application of gene manipulation to aquaculture. *Aquaculture* **85**, 1–20.

Myers, J.M. and Hershberger, W.K. (1991) Early growth and survival of heat-shocked and tetraploid-derived triploid rainbow trout (*Oncorhynchus mykiss*). *Aquaculture* **96**, 97–107.

Singh, R.J. (1993) *Plant Cytogenetics*, CRC Press, Boca Raton, FL.

Manipulation of single chromosomes

Friebe, B., Hatchett, J.H., Gill, B.S., *et al.* (1991) Transfer of Hessian fly resistance from rye to wheat via radiation-induced terminal and intercalary chromosomal translocations. *Theor. Appl. Genet.* **83**, 33–40.

Le, H.T. and Armstrong, K.C. (1991) *In situ* hybridization as a rapid means to assess meiotic pairing and detection of alien DNA transfers in interphase cells of wide crosses involving wheat and rye. *Mol. Gen. Genet.* **225**, 33–37.

Riley, R. and Law, C.N. (1984) Chromosome manipulation in plant breeding: progress and prospects, in *Gene Manipulation in Plant Improvement* (ed. J.P. Gustavson), Plenum Press, New York.

Singh, R.J. (1993) *Plant Cytogenetics*, CRC Press, Boca Raton, FL.

Somatic hybrids

Hammatt, N. (1992) Progress in the biotechnology of trees. *World J. Microbiol. Biotechnol.* **8**, 369–377.

Imamura, J, Saul, M.W. and Potrykus, I (1987) X-ray irradiation promoted asymmetric somatic hybridization and molecular analysis of the products. *Theor. Appl. Genet.* **74**, 445–450.

Itoh, K., Iwabuchi, M. and Shimamoto, K. (1991) *In situ* hybridization with species-specific DNA probes gives evidence for asymmetric nature of *Brassica* hybrids obtained by X-ray fusion. *Theor. Appl. Genet.* **81**, 356–362.

Lindsey, K. and Jones, M.G.K. (1989) *Plant biotechnology in agriculture*, Open University Press, Milton Keynes.

Ohgawara, T., Kobayashi, S., Ishii, S., *et al.* (1991) Fertile fruit trees obtained by somatic hybridization: Navel orange (*Citrus sinensis*) and Troyer citrange (*C. sinensis* × *Poncirus trifoliata*). *Theor. Appl. Genet.* **81**, 141–143.

Somaclonal variation

Beaumont, A.R. (1994) *The Genetics and Evolution of Aquatic Organisms*, Chapman & Hall, London.

Evans, D.A. (1989) Somaclonal variation – genetic basis and breeding applications. *Trends Genet.* **5**, 46–50.

Heath-Paguiso, S. and Rappaport, L. (1990) Somaclonal variant UC-T3: the expression of *Fusarium* wilt resistance in progeny arrays of celery *Apium graveolens* L. *Theor. Appl. Genet.* **80**, 390–394.

Karp, A. (1989) Can genetic instability be controlled in plant tissue cultures? *IAPTC News.* **58**, 2–11.

Karp, A. (1991) On the current understanding of somaclonal variation. *Oxford Surveys Plant Mol. Cell. Biol.* **7**, 1–58.

Single-gene transfer

Aldous, P. (1991) Europe approves first transgenic animal patent. *Nature* **353**, 589.

Botterman, J., and Leemans, J. (1988) Engineering herbicide resistance in plants. *Trends Genet.* **4**, 219–222.

Ezzell, C. (1988) First ever animal patent issued in United States. *Nature* **332**, 668.

Ezzell, C. (1989) Transgenic sticky issues. *Nature* **338**, 336.

Gershon, D. (1991) Will milk shake up industry? *Nature* **353**, 7.

Maclean, N., Penman, D. and Zhu, Z. (1987) Introduction of novel genes into fish. *Bio/Technology* **5**, 257–261.

Palmiter, R.D., Brinster, R.L., Hammer, R.E., *et al.* (1982) Dramatic growth of mice that develop from eggs microinjected with metallothionein–growth hormone fusion genes. *Nature* **300**, 611–615.

Trends in Biotechnology (1993) Special issue: Gene Therapy (issue II).

Index